D0261146

Since Silent Spring

Since Silent Spring

FRANK GRAHAM, JR.

HAMISH HAMILTON · LONDON

First published in Great Britain
by Hamish Hamilton Ltd 1970
90 Great Russell Street London W.C.1

SBN 241 01775 0

Printed in Great Britain by
Lowe & Brydone (Printers) Ltd., London

To Charles H. Callison

Acknowledgments

THE PRESENT BOOK is an attempt to place *Silent Spring* in the context of its time and to recount the progress of the pesticide controversy on which it had so great an impact. I am writing for the nonspecialist, although I hope that here and there the professional will come across material that is new to him. I have not gone into great detail either on the early pesticide fiascoes or the original research that caused thinking men and women to call for a more penetrating look at our pesticide policies. This ground already has been covered by Rachel Carson, as well as by three other scientists whose books stand with *Silent Spring* as the most important written about this subject: Robert L. Rudd, *Pesticides and the Living Landscape*, Madison, 1964; Kenneth Mellanby, *Pesticides and Pollution*, London, 1967; and C. J. Briejèr, *Zilveren Sluiers en Verborgen Gevaren*, or *Silvery Veils and Hidden Dangers*, Leiden, 1967.

Unfortunately, the nature of the attacks by the agri-chemical industry on *Silent Spring* makes it necessary to insert a disclaimer at this point. Like Rachel Carson, I am very much aware of the destructiveness of certain insects. I do not ask that we turn the world over to pest insects, but only that we carefully consider the weapons we use against them. I reject emphatically the foolish notion that we must make a decision between "birds and people."

I could not have written this book without the assistance of many kind men and women. Although final responsibility for the book's accuracy is mine, the following people read all or parts of the manuscript and I am very grateful for the helpful suggestions they made: Shirley A. Briggs, Clarence M. Cottam, William H. Drury, Jr., Paul Knight, Donald I. Mount, and Marie Rodell.

I am also very grateful to the following, who contributed advice or assistance while I was assembling the material: Spencer Appolonio, Enrique Beltran, George G. Berg, John J. Biesele, S. C. Billings, John E. Blodgett, Robert C. Boardman, Dale F. Bray, C. J. Briejèr, Virginia Brodine, William L. Brown, Jr., Maria Buchinger, Marshall Burk, Charles H. Callison, C. H. D. Clarke, Roland C. Clement, Peter J. Conder, Stephen Collins, Stanley Cramp, Dorothy Davis, Ruth G. Desmond, Robert C. Dorion, Robert L. Dow, Eugene H. Dustman, Frank E. Egler, Eugene M. Farkas, John L. George, Richard H. Goodwin, E. Raymond Hall, Evelyn Hanna, Alfred L. Hawkes, Joseph J. Hickey, Paul Howard, Lee Jens, Raymond E. Johnson, J. A. Keith, Eugene E. Kenaga, Robert A. Lafleur, Robert E. Light, Hon. Catherine May, H. Mendelssohn, Donald H. Messersmith, John P. Milton, Charles E. Mohr, N. W. Moore, Allen H. Morgan, Darrell Morrow, Raymond Mostek, David O'Meara, Christopher M. Packard, Colin Platt, Hon. Abraham Ribicoff, Mary T. Richards, E. D. Rolbag, Robert L. Rudd, Peter Scott, Ruth Scott, Rt. Hon. Lord Shackleton, William Shawn, Ronald T. Speers, Marjorie Spock, Murray Stein, Edward A. Steinhaus, Lucille F. Stickel, William Stickel, F. J. Trembley, William M. Upholt, the late William Vogt, Sue Bryan White, Carroll M. Williams, George M. Woodwell, Charles F. Wurster, Jr., and Victor J. Yannacone, Jr.

Finally, I want to express my gratitude to Marie Rodell, who is Rachel Carson's literary executor, for making available to me Miss Carson's papers and correspondence; to Ellen Strout, who typed much of the manuscript; and to my wife, Ada, who gave me considerable assistance and encouragement throughout the task of researching and writing this book.

FRANK GRAHAM, JR.

Preface

THE SCENE is the annual meeting of a state conservation organization. Sitting, with various degrees of attention, on camp chairs that have been set up in the meeting room of an aging but newly gilded hotel are the people we expect to find there — a few assistant professors of biology, the owner of a hunting camp, a cluster of ladies from the local garden club, a TV announcer (hoping to tape a brief interview with one of the guest speakers), a doctor, assorted birdwatchers, hikers, schoolteachers, nature photographers, and perhaps a couple of retired businessmen who live in the hotel and who have slipped into the meeting room in the hope of finding an hour or so of free entertainment.

Various papers have been read, applauded, and forgotten. The usual topics have been trotted out for the disapproval of the usual people — strip mining, road building, air pollution, the dredging and filling of wetlands. Then pesticides are mentioned. The audience stirs. Perhaps three quarters of the people in the room do not know a chlorinated hydrocarbon from a suntan lotion, but suddenly audience-participation has become a part of the program. Someone asks a question, someone else makes a comment. And through the ensuing discussion certain words recur with almost rhythmic regularity:

"DDT" . . . "*Silent Spring*" . . . "Food chains" . . . "Rachel Carson."

The abstractions of conservation have been personified. Beauty and pollution, preservation and spoliation, take on a new reality for the audience. (In a convention of farmers or pest control experts the mention of Rachel Carson's name brings a similar quickening of interest, though customarily it is accompanied by self-conscious guffaws.)

But the reaction among the conservationists is not merely emotional, romanticizing the memory of a dear dead lady, or simplifying an issue. Their interest in pesticides probably was first aroused by Rachel Carson and *Silent Spring*, and these names strike a responsive chord. If America ever chooses to adopt a sane, coordinated conservation policy — an *environmental* policy — a great deal of the credit must go to Rachel Carson. She did more than alert the public to a difficult and critical problem. She uncovered and pointed out publicly for the first time, even to many scientists, the facts which link modern contaminants to all parts of the environment. There are no separate environmental problems, Rachel Carson insisted. She synthesized the issue — for the scientists, the public, and the government.

"There is no question," says one government expert on natural resources, "that *Silent Spring* prompted the federal government to take action against water and air pollution — as well as against persistent pesticides — several years before it otherwise would have moved."

If *Silent Spring*, by its influence, literary quality, and worldwide fame (all three of these attributes are readily granted to the book by its critics), has attained the status of a classic, it has not yet enjoyed the inviolability one usually associates with the classics. It remains very much a living document, exposed to all the buffeting of things alive. Brickbats continue to be aimed at it from high places. Consider these words from an official of the Canadian Department of Agriculture at a recent pollution control conference:

> The furor which has arisen in recent years over the possible "pollution" of our environment by pesticides is one which is

> difficult to understand. Much of this furor developed after publication of the book *Silent Spring*. Although this book served its purpose in that it stimulated an appraisal of the adequacy of information on pesticide residues and their persistence in our environment, many of the statements in the book were inaccurate, and many of the conclusions drawn were based on emotion rather than sound scientific logic.

While opinions of this sort are not rare today, it is virtually impossible to find an eminent scientist, possessed of an ecological outlook, who would associate himself with them. Pesticide residues are worldwide. Since we define pollution as contamination of the environment that interferes with life processes, and since evidence linking persistent, fat-soluble pesticides with such interference continues to be recorded even in the most remote areas, we know that pesticide pollution also is worldwide. Moreover, despite the recent restrictions placed on the use of DDT by the U.S. Government, we inject its relatives into the environment in *rising* quantities each year.*

It has become increasingly important since Rachel Carson's death in 1964 that the environmental issues she raised be kept before the public. That they are kept alive for the readers of scientific and conservation publications is all well and good. But the ultimate decisions about our future will be made by a much wider public. Here, as elsewhere, these decisions are too important to be left entirely in the hands of the experts. Dr. René Dubos of the Rockefeller University has drawn a precise analogy between the microbe-plagued world of a hundred years ago and our own time.

> During the second half of the Nineteenth Century [Dubos writes], it was public pressure organized by enlightened laymen that placed environmental problems at the forefront of scientific

* A *pesticide* is an agent used to kill pests. Under this broad category we find such terms as *insecticides, rodenticides, fungicides, miticides,* and *herbicides,* which are agents used specifically to kill insects, rodents, fungi, mites, and plants.

endeavor. I believe that the situation is exactly the same now. We shall soon experience an environmental collapse unless a grass roots movement makes it imperative that public bodies and the scientific establishment give high priority to the study and control of the forces that are rapidly making the earth a place unfit for human life.

The drama has already begun. The chemist has concocted his compounds, the government has certified them, and the farmer, the forester — and, yes, the home gardener, have spread them abroad. But for good or ill? And if for both good *and* ill, is there a basis for weighing the one against the other — the benefit versus the cost?

"Despite their origin in scientific knowledge and technological achievements (and failures)," writes Barry Commoner of Washington University in St. Louis, "the issues created by the advance of science can only be resolved by moral judgment and political choice."

It will go better for the world when all of us understand that science and technology have, *per se*, no truck with moral judgments. The choices must be made by an informed public. An informed public, in turn, is at the mercy of what ecologist Frank E. Egler has called "the flow of sound and unsound scientific knowledge." The unsound knowledge volleyed insistently at the public today is the result in part of purposeful distortion, and in part of the "specialization" by experts who have fragmented a problem that makes no sense unless it is considered as a complex whole: the problem of maintaining a healthy environment.

As a reporter and a conservationist I see the recent history of pesticide practices and regulation as a problem in communications. The publication of *Silent Spring* in 1962 marked the end of closed debate in this field. Rachel Carson uncovered the hiding places of facts that should have been disclosed to the public long before; she broke the information barrier. Much of the subsequent history of pesticide policy is a response (pro and con) to Rachel Carson's judgment. (Her critics seem to have labored as effectively as her friends to keep *Silent Spring*'s influence alive.) The present book will attempt to describe the genesis of *Silent Spring*, the uproar that followed its publication, and the various books, documents, propaganda pieces, government reports, research papers, and other manifestations of the "flow of sound and unsound knowledge" that have influenced pesticide policy in our time.

Since the impact of Rachel Carson is apparent on almost every phase of this policy, no study of it would be complete without the inclusion of some background material on the woman herself. I have made no attempt to write a formal biography, or even a memoir. I have, however, woven some essential biographical detail through the early chapters to help the reader make sense of a remarkable woman and a remarkable book.

It is not too soon to introduce a note of caution: most of Rachel Carson's critics built their cases on shaky foundations because they did not understand her thesis (in some instances they had not even read her book!). *Silent Spring* is not as simplistic as they had imagined. Rachel Carson did not campaign against the use of all chemical pesticides; and she did not feel that the misuse of certain pesticides was the sole cause of many of the ills she described. These are important points. Chemical pesticides will be with us for a long time to come. The list of their accomplishments is long, and it would be unwise to attribute all symptoms of environmental sickness — whether it is the disappearance of the bald eagle from a part of its range, or the rising incidence of cancer among human beings — solely to the omnipresence of pesticide residues.

Nor is the controversy simply one between "good guys" and "bad guys." Many dedicated public servants and many decent businessmen fail to see what the furor is all about. Like one of Henry James' characters, they have "no perception of difficulties and consequently no curiosity about remedies."

The controversy over *Silent Spring* (and ultimately over all conservation policies) arises from a difference in values. There is no doubt that Rachel Carson's view of life determined the substance of her argument in *Silent Spring*. But her critics' views of life determine the substance of their arguments too. "A soul can be known by its satisfactions," someone has said. Many of us cling to the qualities reflected by Rachel Carson's soul, rather than to those of others whose every decision is based on economics.

Finally, it is well to remember that *Silent Spring* is not the final word. The warnings it sounded remain valid, while the questions it posed remain uncomfortably open.

"There are a great number of questions yet unanswered about the environment we all live in," writes Dr. Virginia Apgar of the National Foundation-March of Dimes, "and there will be more as new drugs, new chemicals such as pesticides and fungicides, increasing pollution and changing social mores continue to alter our environment. The four P's — pills, pesticides, psychedelics and pollution — pose questions of environmentally-caused birth defects we have only begun to suspect should be asked."

The problems are difficult, but the public eventually will have to deal with them. The propagandists who tell us there are no pesticide problems at all, or there are no pesticide problems which cannot be cured by "careful application" of these sophisticated compounds, only confuse the public. They are pushing into the future the decisions we should be making today.

Contents

Part IV: The Uncertain Defenders

Part V: A Light at the End of the Road

Part I

The Controversy

I. The Genesis

> *Some spiritual instinct has shaken itself free and has refused to take the scientific vision of nature as complete . . . It is Miss Carson's particular gift to be able to blend scientific knowledge with the spirit of poetic awareness, thus restoring to us a true sense of the world.*
>
> HENRY BESTON

IN THE COMPOSITION of *Silent Spring* there occurred one of those rare felicitous combinations of subject and author which produce a work of literary interest as well as an impact on a great social debate. Mrs. Stowe's *Uncle Tom's Cabin* is another example. Rachel Carson's subject — the unrestricted proliferation of chemical pesticides in the environment — had for some time caused serious anxiety among certain segments of the public, but factual details remained scanty. The writer herself was uniquely equipped to deal with both the subject and its details. She was a woman who possessed a rigorous scientific background, considerable literary fame, high standards of personal integrity, and a highly poetic sensibility. "A nun of nature, a votary of all outdoors," she has been called. Her closest friends dispute the aptness of that description, feeling that it implies a "cultish" quality about this hardheaded woman. They dispute, too, the comparisons with Emily Dickinson, pointing to Rachel Carson's "rather mischievous sense of humor," and her success in dealing with the worlds of business, government, and academia when she felt it necessary. Nevertheless, there were certain similarities between

Rachel Carson and that other "nun" of American literature: among them a profound relation to the natural world, and an unassailable sense of privacy.

Rachel Carson's life was sharply disciplined, yet there were more than the ordinary number of breathing holes for the human spirit. Her childhood, outside the Allegheny River town of Springdale, Pennsylvania, apparently was lonely. She had been born there, the youngest child by some years of Robert Warden Carson and the former Maria Frazier McLean, on May 27, 1907. The woods around her family's home were the scene of long walks when she was a child, often with her mother, who was a gentle woman with a deep regard for wild things and who remained her companion for most of her life.

Western Pennsylvania is a long way from any coastline. But during her girlhood Rachel Carson was strongly attracted to the great oceans which she had never seen. She especially admired the poems of John Masefield, and vicariously grew absorbed in "the flung spray and the blown spume, and the sea-gulls crying." When she entered the Pennsylvania State College for Women it was with the intention of becoming a writer.

Then, in her sophomore year, she took the required biology course. All of her interests in the outdoors — the woods, the sea, the diversity of living things — fused in this subject. Through her junior year she took more courses in the sciences; in her senior year, having come to a decision, she crammed in the courses needed for a biology major, and began the study of German in preparation for graduate work. She went on to Johns Hopkins where she studied genetics under H. S. Jennings and Raymond Pearl, and later to the Marine Biological Laboratory at Woods Hole, Massachusetts, where she saw the sea for the first time.

Rachel Carson's life seemed neatly laid out for her. The master's degree she had earned (her several doctorates were honorary) led directly into teaching in the early 1930's, first at Johns Hopkins and then at the University of Maryland. But those were difficult years. Depression salaries were low. In 1936 she took an examination for the civil service, and accepted a position with the United States Bureau of Fisheries (which later became a part of the Fish and Wildlife Service). Always prudent, she continued to teach occasional evening courses in biology at the University of Maryland.

And yet, despite her schedule, she had never quite given up her desire to write. "Eventually it dawned on me," she once said, "that by becoming a biologist I had given myself something to write about." Several of her shorter articles were accepted by the Baltimore *Sun*. In 1937 she wrote an introduction to a proposed series of radio programs on undersea life for the Bureau of Fisheries. Her superiors thought it very fine and urged her to expand it into a magazine article. This she did, submitting it to the *Atlantic* and thereby determining her future. "Undersea" was a rare blend of science and poetry. Among its readers were influential writers like Hendrik Willem van Loon and editors like Quincy Howe. Van Loon wrote to her, urging her to write a book about the sea. More substantial — what she later called "that very exciting and flattering letter" — was a suggestion from Howe that she consider writing a book for him at Simon and Schuster.

All her life seemed to have been a preparation for that task. She had read and admired Masefield, Conrad, and Melville. Among her favorite books were H. M. Tomlinson's *The Sea and the Jungle* and Henry Beston's *The Outermost House*. By day she labored at her government job, and in the evenings at her book. The research fascinated her; she preferred to go to the original sources for her material, a practice she followed throughout her life. Often she put off the writing, as she probed deeper into the research material. "I am a slow writer, enjoying the stimulating pursuit of research far more than the drudgery of turning out manuscript." If writing came painfully for her, it did not mean she was less of a writer than a scientist. She shared with that other writer of the sea, Conrad, the sense of despair before the blank sheet of paper, the consciousness of dragging "the ball and chain of one's selfhood to the end."

Late in 1941 her book, *Under the Sea Wind*, was published. Its appearance was accompanied by considerable critical acclaim, the reviewers detecting at once those qualities which later made her famous.

"It has many faults," she said years later about *Under the Sea Wind*, "some of which greater knowledge and maturity may have corrected, but I doubt that a writer ever quite recaptures the freshness of a first book."

One week after her book's appearance, the Japanese attacked Pearl Harbor. *Under the Sea Wind*, like much else in those days, was swallowed up in the ferment as America prepared to fight a global war. It came as a considerable satisfaction to her eleven years later when the book was given a second chance — an event which has gladdened few authors during their lifetimes.

Rachel Carson spent the war years in her office in Washington and later in Chicago, where many of the biologists were shunted to make room for wartime bureaucrats. She was a writer and editor and finally editor-in-chief of the Fish and Wildlife Service's publications. Even if they were obscured by the more spectacular explosions of the war, important events were taking place in her own field — in biology and especially in oceanography. The advances of science in the hothouse of war might reasonably be compared to the pathological growth of living things under abnormal conditions; fertility sometimes grades into monstrosity. Discoveries that might have taken men decades to come upon in peacetime were telescoped into a few short years during World War II. Hundreds of reports, many of them classified under wartime regulations, passed over her desk. This knowledge about the world's oceans, she knew, would have to be interpreted after the war.

She reviewed information of a more disturbing nature too — news of the dramatic accomplishments of certain chemicals which were closely related to the deadly nerve gases. One,

DDT, had performed miracles in smothering typhus epidemics among both soldiers and civilians in Italy. Yet, as a biologist, Rachel Carson harbored doubts about this "savior of mankind" and its ultimate effects upon living creatures. But on this subject, as on the others, there was little that could be written while the war lasted and data were incomplete.

On weekends she and Shirley Briggs, a co-worker at the Fish and Wildlife Service, would pile into the Carsons' aging car and scour the countryside outside Washington for birds. The inevitable collapse of the car threatened for a time to put an end to their excursions. However, no one with an interest in watching birds in wartime Washington needed to give up his hobby for want of a helping hand. The city was alive with experienced "birders" who had been brought together by their wartime jobs — men like Roger Tory Peterson, Irston R. Barnes, and Louis J. Halle (whose delightful book, *Spring In Washington*, is a personal record of the unfolding of the city's natural wonders during that period). The Audubon Naturalist Society of the Central Atlantic States experienced a renaissance. Rachel Carson and Shirley Briggs became part of this group, and formed many lasting friendships while sharpening their skills in the field.

After the war, Rachel Carson continued in her job of editing the Fish and Wildlife publications. "She was a stickler for both accuracy and readable prose, and she was furious at some of the stuff lesser writers turned in to her," a friend has said. "But she never lost that outer calm — she was absolutely unflappable."

Meanwhile, she was making a personal collection of the material gathered by wartime exploration in all the oceans of the world. The idea of writing another book had taken hold of her. There was a great body of knowledge now, and she worked slowly.

"I was years learning to compose on the typewriter," she later said of those days, "and even now when the words are not coming easily I do a difficult passage in longhand. Sometimes I would write most of the night, and sleep in the morning."

In 1950 the separate chapters of her book were submitted to various magazines for first serial sale. One chapter, entitled "Birth of an Island," was published first in *Atlantic Naturalist* and later in the *Yale Review*, and won that year's George Westinghouse Science Writing Award. *The New Yorker* bought nine chapters, condensed them, and published them as a three-part "Profile of the Sea" in June, 1951. But even the early voices of acclaim for these chapters could not have prepared her (or the public) for the vast success of her book. When *The Sea Around Us* was published in 1951 by Oxford University Press, Rachel Carson became a literary celebrity.

"Great poets from Homer down to Masefield have tried to evoke the deep mystery and endless fascination of the ocean," the *New York Times* said of her book. "But the slender, gentle Miss Carson seems to have the best of it. Once or twice in a generation does the world get a physical scientist with literary genius. Miss Carson has written a classic in *The Sea Around Us*."

Her book was chosen as an alternate selection by the Book-of-the-Month-Club; it was condensed in the *Reader's Digest*; it was translated into thirty-three foreign languages ("The books make a beautiful shelf," she said. "Those that are printed in some of the oriental languages look like pages of tapestries.") In this country *The Sea Around Us* remained on the best-seller lists for eighty-six weeks.

At the dinner in New York when she accepted the 1951 National Book Award for Non-Fiction, she alluded to the compliments that had been showered on her prose. "If there is poetry in my book about the sea, it is not because I deliberately put it there, but because no one could write truthfully about the sea and leave out the poetry."

On the heels of this triumph came another. *Under the Sea Wind*, that lovely book that had been swallowed up by the war, was re-issued in 1952 and immediately joined *The Sea Around Us* on the best-seller lists, an event which the *New York*

Times called a "publishing phenomenon rare as a total solar eclipse."

Before her book was published she had taken a leave of absence from her position with the Fish and Wildlife Service. A Guggenheim Fellowship, for which she had applied in order to pursue her studies in marine biology, had removed the last financial obstacle to her escape from office routine. When her book became a best seller she returned the fellowship. But her new fame brought other pressures that terrified this woman who was so intensely jealous of her privacy. "The longer *The Sea Around Us* stays at the top of the list," she wrote to a friend at this time, "the greater, it seems, become the pressures of correspondence, telephone calls and interruptions of all sorts. Even just the labor of saying 'No' in a way that doesn't make people mad takes a good deal of nervous energy and one cannot say no to everything."

As her books finally ran their course on the best-seller lists and the literary world turned its attention to newer stars, Rachel Carson gradually was able to return to work. Writing did not come any easier now that it was her sole profession. She wrote more often by daylight in the house outside Washington that she shared with her ailing mother. Rachel Carson had never married ("I didn't have time," she once said, too polite to say bluntly that it was none of the interviewer's business). She lived, as always, an ascetic life. She took no interest in radio or television. In the evenings after writing she turned to her favorite books, most of which she kept near her bed — Thoreau's *Journal*, Richard Jefferies' nature essays, Melville, Tomlinson, and Beston. During the spring, at the peak of the bird migration, she often arose early to visit nearby wooded areas and watch the vari-colored wood warblers as they passed on the way to their nesting grounds in the north.

Her life, aside from her contacts with a few close friends and relatives, was in her work. Houghton Mifflin Company asked her to write a book about the shoreline (and the life found there) of the eastern United States, and this book became *The Edge of the Sea.* To assemble her material, she spent a great deal of time along the tide lines of the Atlantic coast, from Maine all the way to the tip of Florida. She was able to set off on adventurous expeditions of the kind she had not known since her earlier years with the Fish and Wildlife Service when she spent much time on small boats; in Florida she delighted in exploring the floor of the ocean with the use of primitive diving equipment. Only her domestic duties, in fact, kept her from traveling more widely.

"In minor ways," she once admitted, "I am a disappointment to my friends who expect me to be completely nautical. I swim indifferently well, am only mildly enthusiastic about seafoods, and do not keep tropical fish as pets."

Through her books, Rachel Carson seemed to "own" a little bit of all the world's shorelines, but it was not until 1953 that such a tangible strip became her own. She bought land on the Maine coast, overlooking the tidal pools that had absorbed her attention for so long. This acquisition not only gave her a second home, but it put her in touch with new friends and experiences during the remaining decade of her life. Such a letter as the following, written in 1954 to Henry Beston (whose works she admired but whom she had never met) reveal her diffidence, her graciousness, and her satisfactions. The letter, she told Beston, was at least twenty years overdue:

> It was about that long ago that I discovered *The Outermost House* in a corner of the Pratt Library in Baltimore. I hesitate to guess how many times I have read the book since then, but I don't hesitate to say that I can think of few others that have given me such deep and lasting pleasure, or to which I can return with such assurance of a renewal of my original enjoyment.
>
> While I was writing *Under the Sea Wind* I spent part of a summer at Woods Hole, and one day drove to Eastham and walked down to the beach to find the little house and the surroundings with which I felt so familiar through your pages. After that happy experience I wanted to write you, but procrastinated. Then, a

year or more ago, some one sent me a clipping of your review of *Under the Sea Wind.* (I had missed the review on its appearance.) Let me tell you now, though belatedly, that I found it the most beautiful, perceptive, and deeply satisfying one I had read, and because of my feeling about *The Outermost House*, I was so grateful it was you who had written it.

Recently I acquired a bit of Maine coast at Southport and built a cottage overlooking Sheepscot Bay; last summer we occupied it for the first time — my mother and I. Next week I am going up to open the cottage and spend a few days there. When I was in New York earlier this week, I called Carl Buchheister [of the National Audubon Society] to ask whether veeries would be singing this early, and where. He immediately suggested that I ask you about localities. To me the song of the veery is one of the most deeply moving of bird voices, and I should love to hear them again in Maine.

Writing slowly, going to the original sources, Rachel Carson approached the end of her book on the watery kingdoms and their creatures that make up the Atlantic coastline of the United States. Yet, nearly four years after she had begun her research on the book, she was still grappling with problems of style. In early 1955 she wrote to her editor:

"It needs rewriting. This is true especially of the passages on the reef flat — it doesn't harmonize with the rest of the book — I am still struggling too hard to impart information. I plan to rewrite that part in terms of my own discovery or exploration of the area, or else if impersonally, letting the lives of the creatures flow along in their various interrelationships."

The Edge of the Sea was published later in 1955, after parts of it appeared in *The New Yorker.* The reviews confirmed her stature as one of America's finest writers. Again there were interruptions, demands on her time and pressures that jarred her privacy. Among the pressures were requests from manufacturers who wanted her to lend her name to the promotion of their products; she refused, to the end of her life, to grant such endorsements. Much easier and pleasanter for her were requests from her friends, to which she unfailingly responded. To Edwin Way Teale, who had asked her for help with some obscure water plants he had mentioned in a manuscript, she replied with a long discussion of some minor botanical point. Then, breaking off, she remarked: ". . . and here I go again, talking like the author of *The Edge of the Sea*."

Gratifying, too, were letters from total strangers who expressed their gratitude for the pleasure she had given them. Few writers in our time have had the quality of capturing the hearts, as well as the minds, of their readers quite so completely as Rachel Carson. One woman wrote to offer her the use of a beach house on the Gulf of Mexico.

"I assure you it is from the heart," this woman wrote. "I would just like for you to have time to study the beach and find things that I have never been able to find."

The next couple of years were ones of transition — intensely personal ones, in a way, in which even her work faded into the background as she lived more fully than she ever had before. She and her mother moved into a new home on a large wooded lot in Silver Spring, Maryland, twenty miles from downtown Washington. She built the house to her own specifications, "not being able to find anything that gave me all the special things I need." She also built an addition to the cottage in West Southport. She was in demand too as a writer. Editors wanted articles and books. Since the completion of *The Edge of the Sea* she had been planning a book on evolution (". . . what may prove to be the most important book I have written," she told a friend). As she turned it over in her mind, the original idea slowly grew into a plan to write a book on the ecology of man. But the book, which was to have been published by Harper and Brothers, was never written.

In 1956 she wrote an article for the *Woman's Home Companion*, entitled "Help Your Child to Wonder." The article was based on her experiences outdoors with her four-year-old great-nephew, Roger Christie. She was devoted to the child, whose father had died several years before. On long walks through the woods, or on a visit to a pond, she passed on some of her love, as well as some of her knowledge, to her young companion. In 1957, when Roger's mother died, Rachel Carson adopted him, and he came to live with her in the big brick house in Silver Spring.

She was eager to start work on a new book. For a time she considered expanding the ideas sketched out in her *Woman's Home Companion* article; in fact, she would not have taken time to write the article had she not seen it as the seed of a larger work. (She never wrote that book either, but after her death the article was reprinted almost exactly in its original form by Harper and Row as *A Sense of Wonder*.)

Now, as 1958 began, she was troubled by what, as both a scientist and an observant human being, she saw taking place in the world around her. Life itself seemed to her to be threatened, and therefore her ideas about life were crumbling too. She put some of her anxieties down on paper:

> I have been mentally blocked for a long time, first because I didn't know just what it was I wanted to say about life, and also for reasons more difficult to explain. Of course everyone knows by this time that the whole world of science has been revolutionized by events of the past decade or so. I suppose my thinking began to be affected soon after atomic science was firmly established. Some of the thoughts that came were so unattractive to me that I rejected them completely, for the old ideas die hard, especially when they are emotionally as well as intellectually dear to one. It was pleasant to believe, for example, that much of Nature was forever beyond the tampering reach of man: he might level the forests and dam the streams, but the clouds and the rain and the wind were God's. It was comforting to suppose that the stream of life would flow on through time in whatever course that God had appointed for it — without interference by one of the drops of that stream, Man. And to suppose that, however the

physical environment might mold Life, that Life could never assume the power to change drastically — or even destroy — the physical world.

These beliefs have been part of me for as long as I have thought about such things. To have them even vaguely threatened was so shocking that I shut my mind — refused to acknowledge what I couldn't help seeing. But that does no good, and I have now opened my eyes and my mind. I may not like what I see, but it does no good to ignore it, and it's worse than useless to go repeating the old "eternal verities" that are no more eternal than the hills of the poets. So it seems time someone wrote of life in the light of the truth as it now appears to us. And I think that may be the book I am to write — at least suggesting the new ideas, not treating them exhaustively. Probably no one could; certainly I couldn't.

I still feel there is a case to be made for my old belief that as man approaches the "new heaven and the new earth" — or the space-age universe, if you will, he must do so with humility rather than with arrogance. And along with humility, I think there is still a place for wonder.

These thoughts had been with Rachel Carson for a long time. During the war, when she was still working for the Fish and Wildlife Service, she had been aware of the early studies made on DDT to determine its ultimate effects on the environment. Though some biologists had expressed their apprehensions, there was nothing definite to sustain them. What were thought to be DDT's special assets effectively served to mask its defects. Because it does not break down in the environment, but persists in its toxic state for years, and perhaps for decades, it seemed marvelously convenient for the farmer or forester who could get by with only an occasional application of it to his crops. At the same time, it is not as acutely toxic to animals as many other pesticides are, and therefore seemed to be relatively harmless in every way to "non-pest" species.

Since DDT is cheap and easy to make, it became the most

widely used of the new chemical insecticides. Thus it also became the most intensively studied of them all. DDT was hailed, on the basis of this short-term test, as a "savior of mankind," and the rest of the new chemicals rode in on its coattails with clean bills of health.*

Perhaps the first serious doubts about DDT arose with the discovery that the housefly had developed immunity to it. (Now over 150 species of pests are resistant to DDT.) Insects develop resistance to chemicals in various ways, chiefly through the creation in their bodies of detoxifying enzymes. (See *Silent Spring*, pages 222-29.) A number of scientists, including V. B. Wigglesworth in 1945, forecast the serious impact on the environment of persistent pesticides.

Then, in the 1950's, scientists began to uncover disturbing facts about DDT's effects on "the chain of life." Briefly, this had to do with DDT's unfortunate tendency to accumulate in the fatty tissues of wildlife. The classic example occurred at California's Clear Lake in 1957. There the water was found to contain only .02 parts per million of DDD, a pesticide very similar to DDT in its composition and effects. Microscopic plants and animals in the water stored residues at five parts per million. Yet fish, eating large quantities of the microscopic organisms, concentrated these residues to over 2,000 parts per million. Grebes (diving birds closely allied to the loons) which fed on those fish died in great numbers.

* DDT is one of the organic, synthetic insecticides of the chlorinated hydrocarbon (or organochlorine) group, many of which are held in extremely bad repute by biologists, ecologists, and conservationists. A characteristic of these "hard" pesticides is their stability. They do not break down in the environment, but tend to be recycled through food chains (leaf-worm-robin, etc.); being fat-soluble, they accumulate in high concentrations in the tissues of those animals which are at the top of the food chain. Despite intensive study, no one even today is quite sure how the chlorinated hydrocarbons kill, but they are known to affect the central nervous system. Some other members of this group, besides DDT, are aldrin, dieldrin, endrin, chlordane, heptachlor and toxaphene. (See *Silent Spring*, pages 15-28.) Other major groups of the new chemical pesticides include the extremely toxic but less persistent organic phosphates (parathion, malathion, etc.) and the carbamates (carbaryl).

A year later, Roy Barker of the Illinois Natural History Survey at Urbana published his warning about the intricate cycle of events through which the robin is poisoned by eating earthworms, which in turn have picked up DDT in the leaves under elms sprayed for the control of Dutch elm disease. Roland C. Clement, a vice-president of the National Audubon Society, has since commented on the fate of Barker's warning.

"It was denied or disregarded by officialdom," Clement wrote, "largely because those in responsibility did not regard such environmental sequelae of direct concern to them, and considered birds as 'things' of minor consequence instead of recognizing them for what they are, sensitive and responsive indices to the health and quality of the total environment, of which man too is a part."

Rachel Carson, like many other biologists, was aware of these ominous reports. Yet experienced biologists are human beings too; they read, they listen, and yet they are not always stirred to action until the ominous rumble comes closer to home. For Rachel Carson that moment came with a letter from Mrs. Olga Huckins, a friend who had written for the Boston *Post* under the name Olga Owens. Mrs. Huckins and her husband, a civil engineer, maintained a private bird sanctuary — two acres of fenced-in woodland, with a small pond — behind their home in Duxbury, Massachusetts. In 1957, the state sent spray planes over marshes in Plymouth and Barnstable counties as part of a mosquito control project.

But planes spreading deadly pesticides on mosquitoes are no more certain of their targets than are bombers unloading explosives and napalm on "military installations." Innocent bystanders often are hit just as hard. The spray planes repeatedly crisscrossed the Huckins' sanctuary, taking a heavy toll of birds. Outraged, Olga Huckins sent off a letter to the Boston *Herald*, detailing the incident. At the same time, she sent a copy of her letter, with a covering note, to Rachel Carson. Several years later Rachel Carson wrote to her:

> You deserve credit (or blame, according to the point of view) for having brought my attention back to this problem. I think that even you have forgotten, however, that it was not just the copy of your letter to the newspaper but your personal letter to me that started it all. In it you told what had happened and your feelings about the prospect of a new and bigger spraying and begged me to find someone in Washington who could help. It was in the task of finding that "someone" that I realized I must write the book.

Rachel Carson was horrified by the abandon with which chemicals were poured into the environment, and by the flimsy justifications offered by both industry and government for their unrestricted use. It was apparent that the new chemicals posed a major threat to much of the natural world that she held dear. What ultimate effect this massive chemical barrage would exert on human life itself, no one seemed to know. But she saw clearly that man was, more than ever before, approaching the earth not with humility, but with arrogance.

For the moment, she did not consider this special problem as the topic of her next book. She still envisioned the much broader topic of man and ecology as her coming work. But as part of its background she began to assemble evidence of the havoc man was spreading through his environment by his extravagant use of pesticides. Early in 1958 it came to her attention that the *Reader's Digest* was considering an article dealing with the benefits of aerial spraying, particularly in controlling infestations of the gypsy moth.

"If this is true," she wrote to DeWitt Wallace, the magazine's editor-in-chief, "I cannot refrain from calling to your attention the enormous danger — both to wildlife and, more frighteningly, to public health — in these rapidly growing projects for insect control by poisons, especially as widely and randomly distributed by airplanes."

She went on to give Wallace many of the facts on which she

based her warning. The article was not published; but, as we shall see, this was just the beginning of the *Reader's Digest*'s curious ambivalence toward the problem of chemical pesticides.

But Rachel Carson was not yet convinced that she must postpone her other projects to carry on the battle against the misuse of pesticides. She had taken a keen interest in a court case, in which a number of prominent residents of Long Island's Nassau and Suffolk counties, including Archibald B. Roosevelt and Robert Cushman Murphy, the ornithologist, had sued to prevent federal and state officials from spraying their lands with DDT to control the gypsy moth.

The plaintiffs' action eventually was dismissed. No one as yet had performed sufficient research to establish clearly that link between the spray planes and a poisoned environment. When federal and state agriculture officials joined the chemical industry in mounting a powerful attack on their case, the plaintiffs could assemble neither the scientists nor the documentation to support it in court. Yet, from the beginning, Rachel Carson had seen the Long Island case as a classic example of the citizen's struggle to keep his environment clean and healthy. She saw too that the struggle must be more widely publicized; here an expert and respected writer could be of immense help in alerting the public to the dangers it faced.

At first she tried to interest E. B. White in the project. The reasons for her choice were clear. White, on the staff of *The New Yorker*, was one of America's finest writers. More than that, he had many times expressed his outrage at the rape of America's natural resources in a continuing feature printed in *The New Yorker* called "These Precious Days." On February 3, 1958, Rachel Carson wrote to White, outlining the case of the Long Island plaintiffs, and suggesting that the trial might make an interesting article for *The New Yorker*.

"It would delight me beyond measure," she wrote, "if you should be moved to take up your own pen against this nonsense

— though that is far too mild a word! There is an enormous body of fact waiting to support anyone who will speak out to the public — and I shall be happy to supply the references."

White wrote to her from Maine, where he was then living, to say that he could not undertake the assignment, but he suggested that she might consider doing the job herself. He was sure that *The New Yorker's* editor, William Shawn, would want to talk over such an assignment with her. White concluded:

"I think the whole vast subject of pollution, of which this gypsy moth business is just a small part, is of the utmost interest and concern to everybody. It starts in the kitchen and extends to Jupiter and Mars. Always some special group or interest is represented, never the earth itself."

The course of Rachel Carson's remaining years had been determined. This reluctant crusader had had her mind made up for her. No one else seemed to be in sight to take on the job; certainly no one else with her qualifications — her scientific background, her passionate love for the natural world, and her stature in American letters.

Still, she hung back a little, hoping to carry through with her original plans. She talked it over with Marie Rodell, her agent, with Paul Brooks, her editor at Houghton Mifflin, and with William Shawn of *The New Yorker*. "As you know," she said, "I had various other plans. I feel I should do something on this, however. If I can do a magazine article that would also serve as a chapter of a book on this subject, then also perhaps an introduction and some general editorial work, this would probably be all I should undertake."

Everyone involved was enthusiastic about her participation in such a work. She was swept up in their enthusiasm, and finally agreed to write a brief book on the subject, parts of which would appear in *The New Yorker* before its publication by Houghton Mifflin. The emphasis, in her mind, was on the word *brief*. She already had assembled a great deal of material.

She planned to complete her research as quickly as possible, and then write the book just as quickly so that she might get on with her broader projects. "I hope to complete my work on the chemical poisoning of the environment by insecticides during the summer," she wrote to a friend in the spring of 1958.

2. The Synthesis

Rachel Carson set to work on her book at once. For a time she and her editors used the working title *Control of Nature* for the book, but she came to feel very shortly that the title was too inclusive. To help her to meet her optimistic deadline, Houghton Mifflin hired a magazine writer named Edwin Diamond to serve as her research assistant.

Although the search for material led her into unfamiliar regions of science, she discovered that her previous work opened up lines of communication with surprising ease. In April she wrote to C. J. Briejèr, Director of the Dutch Plant Pest Control Service, asking for information about insect immunity. She introduced herself by explaining that "My books — some of them published in Holland — have dealt with marine subjects." And she concluded with her familiar exhortation: "I shall greatly appreciate a reply by air, since my time for completion of the work is limited."

Briejèr replied swiftly, sending her the information for which she had asked. "Your name is well-known in our country," he told her, "and so is your very well-written book, *De Wereldzee* [*The Sea Around Us*]." Scientists like Briejèr were to be enormously helpful to her during the long labor to come.

As summer began, Rachel Carson was growing aware that a most arduous task did, in fact, lie ahead of her. Each piece of information she uncovered seemed to lead to a dozen more. She came to believe that the full horror of the story lay for the most part unguessed at, even by herself. The research, she discovered, was of a highly sophisticated nature, dealing with much of the most recent work in biology and chemistry (and yet life-and-death decisions continued to be made in this area often by illiterate common laborers!). The material required all of the skills of a trained scientist to evaluate, for little of it had as yet been interpreted for the public.

By June it had become apparent that Edwin Diamond was not in a position to make much of a contribution to the research for the book, an editor at Houghton Mifflin has said. Accordingly the magazine articles, newspaper clippings, and extracts from the Congressional Record which he had collected (most of which Rachel Carson already had gathered during her own research) were returned to him and he played no further part in the preparation of the manuscript. (After *Silent Spring* was published, Diamond attacked the book in an article in the *Saturday Evening Post*, entitled "The Myth of the Pesticide Menace.")

It was clear now in Rachel Carson's mind what her book would be. Using science as its base, it must nonetheless transcend those limited confines of the average scientist's mind which had pulled the world into its current morass. It must not degenerate into vaguely mystical ideas, or smack of the emotional arguments of the fringe groups, which she always tried to keep at arm's length. Science must be the foundation for her work, as it always had been in the past, but it must be given another dimension by the sympathy and compassion without which the finest scientists in the world are dehumanized. She knew that her book must *persuade* as well as inform; it must synthesize scientific fact with the most profound sort of propaganda. She knew that she must be able to sway the professional scientist who often is afraid to stick his neck out unless it is for the simplest kind of one-to-one relationship. It was with this in mind that she began to investigate all the alternatives to soaking the environment with massive doses of chemical pesticides. She was particularly interested in biological controls.

"I'm convinced there is a psychological angle in all this," she wrote to a friend, "that people, especially professional men, are uncomfortable about coming out against something, especially if they haven't absolute proof the 'something' is wrong, but only a good suspicion. So they will go along with a program about which they privately have acute misgivings. So I think it is most important to build up the positive alternatives."

Her resolve was strengthened by the letters of support she received from scientists who had learned of her work. From Dr. Briejèr in the Netherlands she received the first of many letters that were to stimulate her through the most trying days ahead.

> My biological work convinced me [Briejèr said], that the One who was declared dead by Nietzsche and silent by Sartre actually is very much alive and speaking to us through all things. At the same time that work taught me that we could not go on without the collaboration of worms which together with nematodes are killed by nematicides . . . I am very grateful that you took up the subject which I consider extremely important. Please consider my name and work unimportant, however. I am only one of the many voices warning a self-complacent science.

Encouragement came too from William L. Brown, Jr., Curator of Insects at Harvard's Museum of Comparative Zoology (soon afterward he moved to the New York State College of Agriculture at Cornell). "We need biologists and not administrators and not squirt-gun people in charge of our control programs," Brown wrote to her. "Just look at the government's own reports on control of the gypsy moth; it seems obvious that no one in authority had a good, thoughtful, overall look at those data."

But there were sobering words, too. From Clarence Cottam, one of America's leading wildlife specialists, came this remark: "I am sure you will render a great public service, although I shall predict that your book will not be the best-seller that *The Sea Around Us* has been."

Among the scientists whom Rachel Carson was in touch with at this time was Robert L. Rudd, a zoologist at the University of California who had published several important papers on pesticides and their effects upon wildlife. Each was surprised to discover that the other was working on a book about pesticides. Rachel Carson wrote to him in the spring of 1958:

"It should cause no concern to either of us, for I learned long ago that it doesn't matter how many people write about the same thing; each will make his own contribution. As a less experienced writer several years ago I very nearly gave up my plan to write *The Sea Around Us*, just because two other books on the sea came out at that time." And then, in reference to a trip that Rudd was planning to New Brunswick with his family, she invited him to stop for a visit with her at West Southport: "You might enjoy my tidepools, too, although yours are more spectacular."

Rudd and his family accepted her invitation, and they enjoyed a visit together during which they discussed their separate projects. Rudd's book was even longer in the preparation than *Silent Spring*; when it appeared in 1964, however, under the title of *Pesticides and the Living Landscape*, it joined *Silent Spring* as one of the two soundest books yet written in America on the subject of chemical pesticides.

Fuel was added to Rachel Carson's concern during 1958 as word spread among conservationists and scientists about the progress of the United States Department of Agriculture's "Fire Ant Program." The fire ant was an insect which had reached the United States apparently from South America soon after World War I. It had spread gradually through most of the Southern states without causing an undue amount of alarm. At worst, it was considered a nuisance. Then, in 1957, USDA launched a massive propaganda campaign, utilizing press releases, newspaper articles, and motion pictures, against the fire ant. It followed up its verbal barrage with a deadly broadside consisting of dieldrin and heptachlor, two of the most potent of the long-lasting chlorinated hydrocarbon insecticides, in a campaign to "eradicate" the fire ant. Rachel Carson, in *Silent Spring* (pages 132-39), was to give a vivid account of the vast sums of money wasted, and the vast numbers of wild creatures slaughtered, in the useless assault on this insect. But even in 1958 the dimensions of the fiasco were becoming apparent. Wildlife management experts and conservationists were appalled.

"I am pressing ahead just as fast as I can," Rachel Carson wrote to a friend, "driven by the knowledge that the book is desperately needed."

She was not "silly" about wild creatures, as some of her detractors have suggested. To the end of her life she remained a fierce opponent of predator control schemes. She kept a cat, knowing that it would do in some of the wild birds and mammals she loved to have around her home. "I regret the loss of a bird," she said, "but I don't resent it." It was the enormity of the destruction, and its incalculable effects, that drove her to finish her book.

Yet pulling her in an opposite direction and inevitably slowing her work was concern for her family. Rachel Carson was not the "votary of nature" described in the Sunday supplements. She was a middle-aged woman dealing with urgent family responsibilities. Roger, her adopted son, had reached school age and needed much of her time. In September, 1958, he entered the first grade. And all through those autumn months after their return from Maine she cared for her ailing mother, to whom she had always been especially close. The older woman was dying. Rachel Carson worked only occasionally now as she drove back and forth to the hospital. In December her mother died.

It was some weeks before Rachel Carson returned to work. She still hoped to finish the book (which had a new title now — *Man Against the Earth*) by the end of the coming summer. Accordingly, she looked forward to its appearance in the bookstores early in 1960. Writing to William Shawn of *The New*

Yorker, she said that if she had planned an earlier finish for the book it "would have been half-baked, at best." And then she added: "I have a comforting feeling that what I shall now be able to achieve is a synthesis of widely-scattered facts, that have not heretofore been considered in relation to each other. It is now possible to build up, step by step, a really damning case against the use of these chemicals as they are now inflicted upon us."

By this time she was in touch with scientists all over the United States and Europe. She had performed an enormous research task, going through thousands of scientific papers and articles. Only a small part of this material was of significant use to her, but among other papers she found clues that sent her digging in new directions. Her wide reading also indicated to her those people whose ideas and knowledge were the soundest, and therefore would be of most help in assembling her material. From her home in Silver Spring she sent out hundreds of letters. Nevertheless, she was reluctant to disclose to more than a handful of people the full extent of her project.

One of those men she relied on was Clarence Cottam. Cottam once had been her superior in the U.S. Fish and Wildlife Service. Now he was the Director of the Welder Wildlife Foundation in Sinton, Texas. He was one of the most widely respected wildlife management experts in the country, and he already had confirmed her fears about the extent of the damage being inflicted on the environment by the profligate application of chemical pesticides. Cottam served not only as a source of valuable information, but also as a confidant for Rachel Carson during the years she worked on *Silent Spring*.

Early in 1959 she expressed some of her concern to Cottam. "As you know, the whole thing is so explosive, and the pressures on the other side so powerful and enormous, that I feel it far wiser to keep my own counsel insofar as I can until I am ready to launch my attack as a whole."

She knew that, by taking up her pen to write honestly about

this problem, she had plunged into a sort of war. Friends already had warned her that she could expect no quarter from the chemical industry, and she expected none. But even she, a former government worker and an experienced scientist, was amazed at the attitudes she encountered among government agencies and even among official medical organizations.

In chemical pesticides, USDA had found a fascinating new toy, which it was arrogantly flaunting at every opportunity. (If the opportunity did not exist, USDA manufactured one, as it had for its discredited Fire Ant Program.) USDA pesticide "eradication" programs have been aptly compared by Roland C. Clement of the National Audubon Society to the overkill policies manifested in the stockpiling of nuclear weapons by the military. "When combined with the chemical industry's productive overcapacity, and the hustling salesmanship of a free enterprise system, this commitment threatens to poison the landscape and to make the farmer increasingly dependent and the consumer well nigh helpless," Clement has said.

To justify its pesticide barrages, USDA manipulated fact and fiction in a most curious manner; as Cottam pointed out, only a few years before this USDA had estimated the annual crop loss to insects at 10 per cent; by the late 1950's it was estimating this loss, despite a decade and a half of intensive assault by chemical pesticides, at 25 per cent! And, while USDA's programs to eradicate certain insects with the use of chemical pesticides were spectacular failures, the genuinely destructive screwworm (a pest on cattle) had been eliminated from the Southeastern states by the sterilization of the male insect — a biological control.

At this time USDA, on all levels, was extremely sensitive; it became defensive about all of its pest control programs. One prominent scientist, Cornell's William L. Brown, had warned Rachel Carson about the difficulty of obtaining information from USDA's Plant Pest Control Division.

"One of the difficulties," he wrote, "is that the government branches dealing with the pertinent research are either inarticulate or bound to various degrees of administrative reticence. It is probably easier for you to find out what is going on in Beltsville* and Washington and Moorestown* than it is for me. They have branded me as a 'troublemaker' and are so cagey with me that it is laughable." (To which Rachel Carson replied, "I am sure that when my book is out I shall be even less popular with the Division than you are.")

Meanwhile, she learned herself how reluctant USDA officials could be to hand out information on their insect control programs to people whose sympathies they could not count on. One such official in Texas, to whom she had written seeking information on the Fire Ant Program, denied that any serious wildlife losses could be attributed to the insecticides in question. Then he added: "Because of your obviously intense interest in this subject, I should appreciate knowing your affiliation in preparing your report."

Reports coming to her from helpful biologists indicated that USDA's suspicion was directed even toward other government agencies. "The Department of Agriculture representatives have consistently downgraded the biologists' findings and discounted any appreciable damage to fish and wildlife," a high official of the U.S. Fish and Wildlife Service said at the time. And an employee of the Texas Fish and Game commission wrote of USDA's local representatives: "I was disgusted with the flippant way they dismissed all problems. However, it is consistent with all other contact I have had with the USDA Pest Control Division this year. They have refused to give me information on plans for new treatment areas, saying they are convinced no significant wildlife damage occurs and therefore no further checking is needed."

One U.S. Fish and Wildlife Service biologist told of an attempt by state entomologists in Georgia to have him removed from

* USDA's Agricultural Research Center is located at Beltsville, Maryland, and its Agricultural Research Service has a research and regulatory facility at Moorestown, New Jersey.

working in that state by writing letters to Senators Richard B. Russell and Herman H. Talmadge. His sin was questioning the Fire Ant Program in the Southeast. He went on to report the cases of three other biologists in Florida and Alabama "who have either been threatened or have had positive action taken against them by control people." And a biologist with the Tennessee Game and Fish Commission wrote to her to relate an incident that occurred when the Division of Plant Industry applied granules of 10 per cent dieldrin at the rate of 30 pounds per acre for the control of the Japanese beetle. The spraying was so indiscriminate that granules covered picnic tables at a park near Knoxville. Parents and their children had to be told by the biologists to wipe off the tables before setting out their food.

The biologist's letter, like so many others that came to Rachel Carson's attention during this period (which might have been called "The McCarthy Era of the Conservation Movement"), ended with a plea for anonymity: "I would appreciate it if this information is not used in such a way as to jeopardize our relationship with the Division of Plant Industry."

But Rachel Carson had friends even within the Department of Agriculture. After her death, one USDA official asked her literary trustee to return the informative letters he had written her during those years, and burnt them without delay. This, in a sense, is regrettable. While we are likely to think of USDA as part of the regulatory arm of government, it is in effect a branch of our food production industry. It would have been comforting for future generations to know that sanity existed here and there within the department during those profligate days.

The Department of the Interior also was dominated by men who saw themselves only as servants (or partners) of powerful businessmen. On the highest levels this attitude took the form of treating our natural resources simply as objects to be dumped on the marketplace; within the Bureau of Sport Fisheries and Wildlife it was reflected in dominance by the predator control people, who applied their poisons and steel traps at the bidding of ranchers and sheepherders. (See Chapter 16.) Critics inside or outside the department were branded "subversives" hostile to the spirit of free enterprise.

Theoretically, the bureau should have carried out extensive investigations during the 1950's into the effects of the new pesticides on fish and wildlife. This was not the case.

"The fisheries people weren't interested," says John L. George, who served as a biologist with the bureau at the time. "They felt that *over*-population was the fisheries problem then, and I think they even welcomed the new pesticides' effects. 'Maybe pesticides will get rid of some fish for us,' was the way one of them phrased it — and only half-jokingly."

The bureau's Patuxent Wildlife Research Center at Laurel, Maryland, which might have been the center of large-scale investigations, instead became inert. Predator control people, concerned only with exterminating coyotes, prairie dogs, and mountain lions, made the decisions and received the funds. Scientists such as James B. DeWitt, Chief of Wildlife Pesticide Studies, struggled along on an annual budget of $52,000.

"We should have been doing food chain studies and other vital work," says John George, then DeWitt's assistant in charge of field studies. "As it was, Dr. DeWitt performed important pen studies on bob-white quail and pheasants that helped to establish the chronic toxicity of persistent pesticides. And I proposed a nation-wide bald eagle study in 1958. But we got neither funds nor attention."

Officially, the medical world did not present a much more inspiring face. Writing to Clarence Cottam in 1959, Rachel Carson referred to some of the American Medical Association's excellent pamphlets on the dangers of pesticides. "The position of the AMA itself seems most ambiguous," she wrote. "I have the impression that, when called on to take a stand, the Association is on the fence, or even somewhat on the wrong side. Personally, I propose to ignore this, and to quote many of their own statements with what ought to be damaging effect."

Her impression was confirmed by a number of physicians with whom she was in contact. One, from Illinois, wrote: "I must agree with you that the medical profession is probably not yet aware of the importance of the use of poisonous substances as insecticides and herbicides. Also, I am afraid that public health officials are not always as cognizant of the importance of these hazardous substances as they should be." This physician went on to deplore a recent bill, aimed at pesticide restriction, that had been passed by the Illinois legislature. "It has now been so thoroughly amended that it becomes essentially a 'labelling' act."

At about this time Rachel Carson received some encouragement from an unexpected direction. In June, 1959, there appeared in the *Reader's Digest* an article strongly critical of current pesticide use called "Backfire in the War Against Insects." The article, which was written by Roving Editor Robert S. Strother, mentioned many of the alarming incidents which had prompted Rachel Carson to go to work on her book. The *Digest*'s editors pointed this out in their note which accompanied the article. It read: "There is mounting evidence that massive aerial spraying of pesticides may do more harm than good. Until the full results are known, all concerned should heed the warning: 'Proceed with Caution.'" (This was the conclusion that Rachel Carson presented, with much more detailed and up-to-date information, three years later when *Silent Spring* was published; but by then the *Reader's Digest* had undergone a curious change of heart.) The *Digest*'s editors received many letters congratulating them for the article, but among the correspondence came letters from two different USDA officials, which referred to "exaggerations" and "misinformation" but avoided answering any of the article's main points.

By now all of Rachel Carson's plans for finishing her book that summer had been dashed. New avenues of research appeared every day. Letters arrived with clues to be tracked down and facts to be checked. In the midst of all this she contracted a streptococcus infection in June, delaying her departure for

Maine. Though she and Roger finally reached West Southport, her illness dragged on through the summer. Her work was slowed. Finally, late in the year, she was able to ease some of her burden by finding just the assistant she had been looking for. Mrs. Jeanne Davis was the wife of a doctor attached to the U.S. Public Health Service. She had worked with medical men at the University of Rochester and Harvard Medical School, and was familiar with technical publications. As Rachel Carson's secretary, Mrs. Davis relieved her of much of the burden of correspondence and thus played a significant role in the increasingly complex task of bringing the book to a conclusion.

Rachel Carson was, in fact, beginning to put together her book at this point, working over small random areas of the story rather than writing a consecutive draft. She had already "finished" the first chapter (although, later, it would be entirely rewritten). Then she skipped to a subject that was of increasing interest to her — the cancer-producing aspects of chemical pesticides. Earlier, she had merely intended to treat this subject as part of a general chapter on "hazards to man." Now she knew that it must take up one, and perhaps several, chapters of her book.

By the year's end she was still far from finished with her book. "I guess all that sustains me," she wrote to her editor, "is a serene inner conviction that when, at last, the book is done, it is going to be built on an unshakable foundation. This is so terribly important. Too many people, with the best of motives — have rushed out statements without adequate support, furnishing the best possible target for the opposition. That we shall not have to worry about."

In the belief that her health troubles were comparatively minor, she worked on during the summer months. But in the fall of 1960 she learned that she had not been told the truth about her condition. A tumor removed by surgery the preceding March had been malignant. She began radiation treatments at once, then flew to a Cleveland clinic for further examinations.

By Christmas she was home again in Silver Spring, from where she wrote to a friend: "Perhaps more than ever, I am eager to get the book done." And then, referring to Christmas Day, she added, "We had a happy day here — who could help having a Merry Christmas with an eight-year-old boy surrounded by space-age toys?"

Though publication seemed farther away than ever, she continued to work on isolated chapters of her book. A chapter on cancer, still in rough form, was sent off to be checked for accuracy by several medical men, including Dr. W. C. Hueper of the National Cancer Institute. She was also winding up chapters on herbicides, and on the effects of mass spraying. She was even concerned with that elusive detail, a title for her book; earlier ones had been discarded, and now she was leaning toward *Dissent in Favor of Man.*

During these difficult months she was sustained not only by the sympathy of her friends and the conviction that she had important work to finish. Every mail seemed to bring some heartening note. From the University of Texas came a letter, written by a scientist whom she knew only by reputation, John J. Biesele, a Professor of Zoology:

> I am overjoyed to learn that you are working on a book on this subject, because your writing ability and public appeal will weigh heavily in the final balance. Something must be done, in my opinion, to counteract the increased "education of the public" recommended by some of the commercial producers of chemical pesticides in an issue of the *Chemical and Engineering News* last fall.

There were also letters containing fascinating pieces of information. Through the U.S. Fish and Wildlife Service in Denver, she learned that marine fish, both mackerel and butterfish, were found to contain significant traces of DDT. She had not yet heard of DDT in marine fish, and she wrote asking for more information.

"These fish were apparently purchased in a fish market," the reply explained. "We suppose that the DDT was picked up by the fish in a packing plant." The full extent to which DDT has permeated our environment, loading even the vital organs of birds and fish in the antarctic, had not yet been suspected!

When Rachel Carson left for Maine in the summer of 1961, she already had been working on her book for over three years. She was desperately anxious to see the book in print, because nothing had changed to allay her fears about the effects of the present use of chemical pesticides. USDA seemed unchastened by its Fire Ant Program, and in fact continued to press for the eradication of that second-rate nuisance. Industry promotion and advertising were increasing. So too were sales and the resultant deadly mist which had enveloped the nation, like the debris from some inexhaustible Krakatoa.

This is not to say that, in some quarters, concern was not increasing too. Doubts about the wisdom of current pesticide practices had spread from a few biologists and conservationists to include many other citizens in and out of government. In his "Special Message on Natural Resources," sent to Congress on February 23, 1961, President Kennedy had called for greater consistency and coordination of leadership "without the present conflicts of agencies and interests." The President then referred to "one agency encouraging chemical pesticides that may harm song birds and game birds whose preservation is encouraged by another agency."

Several months later, John V. Lindsay, then a congressman from New York, wrote to the Secretary of Agriculture, asking if new legislation was needed to combat the pollution of the environment. He received a reply from Assistant Secretary Frank J. Welch that existing law adequately protected the public interest and that additional legislation was unnecessary.

The public's fears also were being aired in the press. There had been some discussion of pesticide use in Maine newspapers. Rachel Carson was reluctant to have her "ammunition," as she called it, given wide exposure before she published it in her book. But in August she was prompted by the articles to write a letter to a local paper, the *Boothbay Register*, emphasizing the idiocy of trying to control such insect-borne diseases as the Dutch elm disease by massive spraying programs. But her letter was picked up by interested citizens and given national attention.

"To write as I did for a small and local paper seemed deceptively harmless, and I should have known better," she said afterward. Accordingly, though she was sorely tempted, she refrained from becoming involved in a controversy about pesticide damage that erupted that fall on the editorial pages of the *New York Times*.

A staphylococcic infection in the winter of 1961-62 set back her work dishearteningly.

"About the only good thing I can see in all this experience," she wrote to a friend, "is that the long time away from close contact with the book may have given me a broader perspective which I've always struggled for, but felt was not achieving. Now I am trying to find ways to write it all more simply and perhaps more briefly and with less exhaustive detail."

This became an increasingly difficult part of her task. The material was overwhelming in volume, and often subtle in detail. How could she get the story across to the great mass of readers, untrained in science, who must ultimately provide the pressure that would bring about saner policies? How could she make chlorinated hydrocarbons compelling? She concluded that it required writing and rewriting, almost endlessly.

3. A Most Curious Affair

"I AM CERTAIN you are rendering a tremendous public service," Clarence Cottam wrote to Rachel Carson early in 1962, as he returned (with some suggestions for corrections and additions) those parts of her manuscript which dealt with wildlife problems. "Yet I want to warn you that I am convinced you are going to be subjected to ridicule and condemnation by a few. Facts will not stand in the way of some confirmed pest control workers and those who are receiving substantial subsidies from pesticide manufacturers."

There was ample basis for Cottam's premonition. At the time he was himself up to his ears in a disquieting affair that also concerned the use of chemical pesticides. Because this episode goes far to explain the reception *Silent Spring* received from professional people after its publication, and indeed to explain the whole of the subsequent pesticide controversy, it is worth our while to pause briefly and look into its background. The affair concerned the prestigious National Academy of Sciences-National Research Council.

The NAS-NRC, a private, nonprofit organization of outstanding scientists, was organized in 1863. It is dedicated to "furthering science for the general welfare," and on request it is required by its charter to act as official advisor to the United States Government on scientific matters. By its tradition, its membership, and its links to the government, the NAS-NRC would seem to be able to speak with authority on whatever topic it undertakes to study.

Cottam was responsible in part for the Academy's plunge into pesticide matters. Like many other biologists, he had been greatly troubled by USDA's Fire Ant Program in the late 1950's. Time and again he heard from colleagues the story of the program's disastrous consequences. Even more disturbing were letters from men whom he knew to be competent biologists, telling him that federal and state agriculture officials were attempting to have them fired or transferred for their opposition to the program. Cottam protested to Ezra Taft Benson, who was then the Secretary of Agriculture.

"I told Benson I was confident that within a few years he and his agency would be severely criticized," Cottam says. "I told him that in my judgment the Fire Ant Control Program was a disgrace to the Department because it was not being conducted scientifically, and I urged that he have an independent agency make an objective study of it."

USDA responded by appointing a biologist, Harlow B. Mills of the Illinois Natural History Survey, to investigate the program's adverse effects upon wildlife. Cottam learned from friends who saw the report that, although Mills was by no means hostile to USDA, he was not enthusiastic about its Fire Ant Program. Cottam wrote to Secretary Benson for a copy of the report.

"I was assured I could see it if I would treat it confidentially," Cottam says. "I said that I did not want to see it under those conditions as I wanted to use the report. The report was filed away and was never permitted to be published because it was derogatory to USDA and its control group."

Nothing had been solved. As public clamor rose against the Fire Ant Program, as well as against several other massive USDA insect "eradication" schemes, Secretary Benson threw the problem into the lap of the National Academy of Sciences. In the summer of 1960 it was decided to establish, within the Academy, a Committee on Pest Control and Wildlife Relationships. The committee was composed of seven members representing industry and several branches of science. Its chairman was I. L. Baldwin, professor of Agricultural Bacteriology at the University of Wisconsin.

The committee in turn appointed three subcommittees, each composed of eight members, to make a factual study of the whole problem of the effects of chemical pesticides on wildlife. (None of the men serving on the committee or its subcommittees was a member of the Academy.) Each subcommittee was asked to submit one part of the proposed three-part study, the project to be paid for by contributions from the chemical industry, conservation organizations, and federal agencies. The assigned topics were:

Part I — *Evaluation of Pesticide-Wildlife Problems*
Part II — *Policy and Procedures for Pest Control*
Part III — *Research Needs*

Two years later (but before Part III of the report even had been finished) committee chairman Baldwin commented on the success of the venture during the course of his unfavorable review of *Silent Spring* in the magazine, *Science*. The subcommittees, he wrote,

> have made a careful and judicial review of all the evidence available, and they have published a series of reports making appropriate recommendations. These reports are not dramatically written, and they were not intended to be best sellers. They are, however, the result of careful study by a wide group of scientists, and they represent balanced judgments in areas in which emotional appeals tend to overbalance sound judgment based on facts.

How accurate was the committee chairman's description of his group's handiwork? Let us look at the reports, and at the comments of biologists who reviewed them. (Since the reports deal with the effects of these chemicals on wildlife, it is assumed that biologists are qualified to make the soundest evaluation of the problem — yet, according to one critic, "agriculturists and entomologists dominated" the subcommittees.)

The whole project might have been pulled off smoothly had not Cottam been named a member of the third subcommittee, investigating "Research Needs." He was not alone, however, in his unhappiness over the course of the investigations. Outsiders heard that there were frequent squabbles in all three subcommittees about the reports' contents and conclusions. It soon became apparent that the results would be not the hoped-for consensus, but a compromise among the several factions.

The difficulty lay in the conflicting outlooks of the subcommittee's members. Many of these men came to the investigation secure in the belief that chemical pesticides were an unalloyed blessing to man and his environment. Others felt that the reason they had come together in the first place was to test that notion. The latter group saw in the appointment of W.H. Larrimer to the position of executive director of the committee a reflection of the committee's initial bias.

"To begin with," Cottam recalls, "Dr. Larrimer was a dominating influence in determining the policies and conclusions. He had spent most of his life in USDA's Insect Control Division, and had worked especially on DDT and other chlorinated hydrocarbons. He was convinced beyond any question that no ill effects could result from the use of any of these chemicals."

Another interesting story was that of George C. Decker, who had been appointed chairman of the Subcommittee on the Evaluation of Pesticide-Wildlife Problems. An economic entomologist with the Illinois Natural History Survey, Decker was a very able scientist. He had served as an advisor and collaborator on insect control to the U.S. Department of Agriculture, and frequently as a consultant to various chemical companies. It was Decker who, in a review of *Silent Spring* for *Chemical World News*, was to say: "I regard it as science fiction, to be read in the same way that the TV program 'Twilight Zone' is to be watched."

Yet Cottam pointed out the sharp divergence in Decker's point of view in the 1960's with that expressed by him a decade before. In 1950, before the annual meeting of the North Central States Branch of the American Association of Economic Entomologists, Decker had had the following to say:

> Chemical control of insects is only one phase of insect control, yet it appears that the urgent demand for information on new insecticides has led all of us into a large-scale, fadistic swing to insecticidal investigation at the expense of our other research . . . I believe . . . (as) some of our illustrious predecessors (have said) that man, as a rational and intelligent being, should be able to outwit insects and not rely entirely upon chemical warfare . . . Insecticides are extremely useful in protecting crops from immediately impending damage, but have little effect in influencing insect abundance from one year to the next . . . Insecticides are fire-fighting, not prophylactic, weapons . . . (they are) habit forming in that once their use is started their continued use becomes more and more necessary. Although the fruits in small home or other unsprayed orchards are usually wormy, they are seldom as badly infected as the apples on a few check trees in a sprayed orchard or in the commercial orchard the first year it is abandoned . . . About all we have ever obtained from insecticides was annual crop protection . . . It seems quite obvious to me that we should not and cannot consider the use of chemicals a substitute for sound cultural and other biological control methods. We have been amply warned that many of the new insecticides can and often do upset the biological balance in an area and while promoting more effective control of one pest we produce an equally more destructive outbreak of some lesser pest . . . When properly used insecticides are very valuable tools, but like the "A" bomb, if unwisely and wrongly used, they may lead us to our doom. It seems to me we are in the position of the drunk in a high-powered car approaching a stop and go light. We had better sober up, look and listen for danger signs before we proceed much further.

Here was a brilliantly perceptive statement on the use of chemical pesticides. All that had happened in the intervening years was the accumulation of further evidence to sustain that statement. Yet Decker had grown increasingly antagonistic to its point of view.

By 1962, the first two subcommittees had finally agreed on their reports. These reports were to be published under curious "ground rules." It was agreed that nothing was to appear in any of the reports that did not have unanimous approval within the subcommittee concerned. Any fact or conclusion that proved unacceptable to any subcommittee member was expunged from the final report. No minority report was permitted. Despite the internal wrangling, the two reports were ground out approximately on time (although two members of those subcommittees told Cottam that material in the reports was altered after they had put their signatures to them).

Part III of the report became bogged down in controversy. Cottam was joined in vigorous opposition to the pesticide-oriented conclusions by Clarence Tarzwell of the United States Public Health Service. They believed that Larrimer, the executive director, was unduly influencing the tone of the report.

"Larrimer objected so strenuously to my conclusions and comments that he wrote and said he would not accept them," Cottam recalls. "He indicated how many lives had been saved by DDT and how much money had been saved. I described all of his comments as plain unadulterated bunk with no scientific basis. I then appealed to the chairman of the overall committee and asked why a subcommittee was chosen if the executive secretary was going to tell us what our conclusions should be and how the work of the subcommittee should be carried on."

When the first two parts of the report were published in 1962, Cottam was shocked. So were a number of other biologists. Writing in *Audubon Magazine*, Roland C. Clement had this to say about the NAS-NRC Report:

> Given their authoritative backing, these two reports are extremely disappointing . . . The whole is a generalized and undocumented

> statement that, far from coming to grips with the problem, seems to disregard much important evidence and does no more than offer a gentle admonition to pesticides users to be more careful in the future. This result is either a mark of ecological incompetence in the committee, or more likely, evidence that the viewpoints of the advocates of pesticides use within the committee prevailed almost entirely.

And Frank E. Egler, writing in the *Atlantic Naturalist*, was even more scathing:

> These two bulletins cannot be judged as scientific contributions. They are written in the style of a trained public relations official of industry, out to placate some segments of the public that were causing trouble . . . The problem of industries' influence on scientists who are on their payrolls as consultants, through research grants, or otherwise, is a prickly one. It has been brought up in connection with these reports. My surprise is not that such influence exists, but that other scientists are so naive and unsophisticated as to refuse to believe it. The reader should at least know of such connections in appraising the final conclusions.

How, indeed, do these reports strike the interested reader?* The first thing he notices is that, although they purport to be scientific documents, they contain no documentation whatever. The reader might be forgiven for wondering if they were composed by some anonymous drudge in one sitting, wholly in the absence of notes or source material. Throughout, there is no distinction made among the many types and relative toxicities of pesticides, and their varying effects under varying conditions.

To open Part I, *Evaluation of Pesticide-Wildlife Problems*, and turn to the Table of Contents is to receive a lesson in public relations. The first seven pages of this twenty-eight-page booklet are devoted to an Introduction and to the topic "Pesticides A

* The assembled report was published under the title *Pest Control and Wildlife Relationships*. National Academy of Sciences–National Research Council Publications 920-A, 920-B, and 920-C. Washington, 1962, 1963.

Modern Necessity." This section, fully 25 per cent of the booklet, is simply a litany of praise for chemical pesticides. There is no clear statement of what the problem is, nor of the specific problems that prompted the study.

The report tells us that a vast arsenal of new chemicals has been "put to use on an increasingly wide scale to protect man and those plants and animals essential to his welfare." It does not mention the fact that those chemicals have been put on the market without adequate testing in the field. It does not tell us who makes the decisions about when man and his animals need protection, nor who decides on the treatment, the kind of poison to be used and the rates of application. Nor does it tell us how we may draw the line between man and his domestic plants and animals on the one hand, and those other organisms in the environment which may be equally important to his welfare.

The marvels of pesticides are emphasized again and again; their drawbacks, including the evolution of resistant insects, are barely mentioned. Barely mentioned, too, are alternate methods of control; we find statements that the development of these alternate methods did not keep pace with farmers' needs. (The development of biological and cultural methods of control, the report might have added, did not keep pace with the profitable manufacturing and promotion of chemicals.)

In contrast to this lengthy promotion of pesticide wonders, only two pages, plus five lines, are devoted to "Wildlife Values." As Egler wrote in his review, "If this is all that can be said in defense of wildlife, one marvels that this committee was ever set up. Wildlife here seems to be something that annoyingly gets in the way of pest control programs." Not only is the space limited, but the homage to wildlife is banal in the extreme. Consider the following paragraph which plays an important role in this "scholarly" evaluation of wildlife in America.

"A survey of sport-fishing and hunting activities of people 12 years of age and over showed that more than 30 million Americans spent nearly four billion dollars in some 658 million days of hunting and fishing in 1960. These sportsmen came from 20 million households. Moreover, there are countless people who enjoy wildlife for its aesthetic value and as a source of relaxation from the pressures and tensions of modern living."

If the presentation here is any indication of enthusiasm, the reader must conclude that the hearts of the authors are attuned to praising pesticides rather than wildlife!

The last section of the booklet is concerned with "The Impact of Pesticides on Wildlife." Here is the booklet's *raison d'être* and an opportunity to wipe out its earlier shortcomings. With rising hopes the reader quickly comes across the following statement: "With the thousands of tons of pesticides that are used each year in the control of pests, it may be concluded that, to date, the impact on wildlife, although not disastrous, is just cause for concern."

But the impact is just as quickly dissipated. Pussyfooting abounds. Such soft spots as "may occur," "may expose," "seems to indicate," "under certain conditions," etc., undermine the statement of pesticide effects. Moreover, even when the authors focus on hazards to specific forms of wildlife, the absence of information about dosages, seasons and numbers of individuals affected renders the statement almost meaningless. The approach is simplistic, as if the authors had no comprehension of the complexities involved in large-scale attempts to manipulate the environment. The section closes without a single mention of the effect of pesticides, particularly herbicides, on wildlife habitat.

A sketchy section, "Summary and Conclusions," is appended to the report. Again the reader hears that chemical pesticides are essential; damage to wildlife is "minimal"; and when such damage occurs, it is attributed to pesticide "misuse." Although the report had urged several times earlier that additional research be undertaken, there is no mention here either of research into the effects of chemicals on wildlife, or into alternate means of pest control.

On the surface, Part II, *Policy and Procedures for Pest Control*, seems to be a more professional job than its predecessor. It is almost twice as long (fifty-three pages) and goes into greater detail. There are sensible statements on research needs and on the case for cooperation among various state and federal agencies to protect wildlife from pesticide hazards.

Yet the report is, in a real sense, a whitewash because it clings to the notion of "things as they are." It shows no urgent desire to upset the present structure, or to move in the direction of alternate controls. The fact that the entire study was prompted by USDA's disastrous policies and procedures in its massive eradication programs is studiously ignored. Instead, the report suggests that programs directed by federal and state officials are above reproach. Our "minimal" difficulties may be traced to the little fellow who fails to read the label.

Here is Frank E. Egler's criticism of the report:

> There are many "musts" and "shoulds." The bulletin closes with a set of seven "Precautions" all starting with the words "Do not apply . . ." unless, or until, or with some other such qualification. Perhaps I am inordinately naive, but it had been my simple impression that the entire pesticide problem, leading to the formation of the Committee itself and to the "study" resulting in these publications, was that in actual practice these precautions have not been and could *not* be followed. In short, now that the "precautions" are stated, the work of the Committee can begin!

After the publication of the report's first two parts, Cottam and Tarzwell struggled more persistently than ever to produce a balanced study. There is no doubt that the committee itself was embarrassed by the delay in producing Part III. The executive secretary asked Cottam to resign, but no further steps were taken against him when he threatened to publish his own minority report. The dispute came to the notice of the highest officials of the National Academy of Sciences. A member of the executive board of NAS-NRC visited Cottam at his home in Texas, urging him to stick to his guns if he felt strongly about the nature of the report.

A report of sorts finally was pasted together. It was released as Part III, *Research Needs*, a year and a half after its fellows, and nearly a year after the publication of *Silent Spring*. Even so, Cottam claims that statements of which he does not approve were inserted in the report after he had signed it.

The reader may confine his investigation of Part III to the Introduction. It is candid indeed: "Originally it was intended to review and evaluate the technical information available on pest control and wildlife relationships, the research being done, and the research needs. However, in the interest of brevity it was decided to concentrate on the development of a research program to meet future requirements."

What began as a potentially controversial but valuable report had been emasculated. Cottam, himself, has given the last word on Part III:

"It isn't worth the paper it's written on!" he wrote to a friend.

I have taken the time to review the background of this report, not because of its quality, but because of its reputation as an important scientific study. It was quoted as such by chemical trade associations, especially following the publication of *Silent Spring*. Men with philosophies similar to many of these subcommittee members "savaged" Rachel Carson at that time. It is interesting to note that early in the struggle to produce a balanced report, Cottam was criticized by one of the members of another subcommittee, Dr. Mitchell R. Zavon. Zavon, Associate Professor of Industrial Medicine at Kettering Institute and a consultant to the Shell Chemical Company, replied to Cottam's fears that pesticides were unfavorably altering our environment.

"We alter nature daily," Zavon said. "Would you have us believe that you or I or anyone else really knows what is nature's way? We are natural creatures, and who is to judge whether or not our destructiveness, however we may deplore it, is not an ordained path in nature's road of terrestrial development."

Ironically, it was those scientists sharing Zavon's viewpoint who later attacked Rachel Carson for "mysticism" and "unscientific statements!" Little wonder then, in view of his experience with the members of the various NAS-NRC subcommittees, that Cottam was apprehensive about the treatment a segment of the scientific world would deal *Silent Spring*.

*

Throughout the spring of 1962 Rachel Carson worked almost without pause to finish her book. She was still rewriting chapters, in part or in their entirety. By now it was common knowledge in many quarters that she was working on an "important" book. Supreme Court Justice William O. Douglas wrote to her, asking when it would be finished.

A title had at last been found. Her editor had first suggested *Silent Spring* as the title for the chapter on birds; and later, as its broader meaning became apparent, for the book as a whole.

For a time *The New Yorker*, which planned to begin printing portions of the book in June, considered running it in five long installments; editor William Shawn was so impressed by the book that he wanted to cut as little of the text as possible. But Rachel Carson objected. She believed that it would be too much to ask of a reading audience, most of it untrained in the sciences, to follow the thread of a complicated argument over a period of five weeks.

It was decided accordingly to publish the condensed version of *Silent Spring* in three parts in *The New Yorker*, beginning on June 16, with the second and third parts appearing in the next two weekly issues. About one third of the book's total text would appear in this manner. The book itself would be published by Houghton Mifflin late in September.

4. The Counterattack

SILENT SPRING IS NOW NOISY SUMMER. This head appeared over a story in the *New York Times* on July 22, 1962. *Silent Spring* was not yet between hard covers, but the uproar in government, chemical, and agricultural circles was intense. The serialized and abbreviated version of Rachel Carson's book in *The New Yorker* had created a greater stir than anyone earlier had imagined.

The reaction was as varied as it was intense. Friends wrote to Rachel Carson from as far away as the mountains of the Northwest and the villages of the Maritime Provinces that the chief topic of conversation seemed to be *Silent Spring* and chemical pesticides. The Toledo, Ohio, Public Library ordered a large supply of ladybugs from California (at $6.50 a gallon) and released them on its grounds to control aphids. In Bethlehem, Pennsylvania, the *Globe-Times* described the reactions of farm bureau personnel in two nearby counties: "No one in either county farm office who was talked to today had read the book, but all disapproved of it heartily."

Some chemical firms reportedly instructed their scientists to examine the articles line by line, probing for weak spots. The National Agricultural Chemicals Association preferred not to meet Rachel Carson head on. Instead, this powerful lobbyist for the chemical manufacturers expanded its public relations program and published a number of new brochures which reaffirmed the unadulterated blessings of chemical pesticides.

At least one chemical firm preferred to take direct action. On

August 2, the Velsicol Chemical Corporation of Chicago addressed a five-page registered letter to Houghton Mifflin, suggesting that the company might wish to reconsider its plans to publish *Silent Spring*, especially in view of the book's "inaccurate and disparaging statements" about chlordane and heptachlor, two chlorinated hydrocarbon pesticides manufactured solely by Velsicol. The letter was signed by Louis A. McLean, Secretary and General Counsel of Velsicol. The letter's sentiments reached their climax in the following paragraph:

> Unfortunately, in addition to the sincere opinions by natural food faddists, Audubon groups and others, members of the chemical industry in this country and in western Europe must deal with sinister influences, whose attacks on the chemical industry have a dual purpose: (1) to create the false impression that all business is grasping and immoral, and (2) to reduce the use of agricultural chemicals in this country and in the countries of western Europe, so that our supply of food will be reduced to east-curtain parity. Many innocent groups are financed and led into attacks on the chemical industry by these sinister parties.

Houghton Mifflin asked for more detailed information on the statements in *Silent Spring* to which Velsicol objected. On receiving the information from McLean, Houghton Mifflin asked an independent toxicologist to review the disputed material. When the toxicologist confirmed the accuracy of Rachel Carson's statements, her publishers informed Velsicol that the book would appear as planned, and nothing further was heard about the matter.

There was furious activity behind the scenes at all levels of government. The Federal Pest Control Review Board (ineffective precursor of the present Federal Committee on Pest Control) met in a troubled session. An observer at that meeting recently recalled its distasteful tone.

"The comments alternated between angry attacks on *Silent Spring* and nasty remarks about Miss Carson," this government official said. "One well-known board member, I recall, said, 'I thought she was a spinster. What's she so worried about genetics for?' Some of the other board members thought this was very funny. I was disgusted by the whole meeting."

Elsewhere the book was taken more seriously. The *Wall Street Journal* had this to say in its August 3 edition about the Department of Agriculture's concern: "Secretary Freeman squelches trigger-happy underlings who itch for a quick rebuttal of Rachel Carson's magazine attacks on safety of chemical insecticides. The Agriculture Department builds a careful defense of its encouragement of insecticide use. An indirect reply: The Department pushes work on non-chemical war against bugs."

Over at the Department of the Interior, Secretary Stewart L. Udall also grasped the significance of the book. He realized that the problem it discussed soon would become a major issue, and he wanted to be able to take an intelligent stand whenever it was called for. Accordingly, he assigned one of his assistants to follow every phase of the book's publication, and the controversy that followed it. (It is interesting to note that an intelligent stand was called for the following year when Udall testified before Senator Abraham Ribicoff's subcommittee which was investigating pesticide use; at that time, Udall became probably the first prominent public servant to point out that we are contaminating the entire environment.)

Congressman John V. Lindsay wrote to Rachel Carson to tell her that he had found *The New Yorker* articles to be "a detailed and persuasive contribution to public awareness of the dangers of our present pest control policy." Lindsay inserted the concluding paragraphs of the first installment in the *Congressional Record*, and told her, "I wish that I could have inserted the entire article."

Part of *Silent Spring*'s impact on the public that summer resulted from the coincidence that *The New Yorker* articles followed almost immediately the thalidomide tragedy, in which it was discovered that this new and inadequately tested chemical tranquilizer, when administered to pregnant women, often caused serious physical abnormalities in their babies. Both the public and government officials, therefore, were receptive to cautionary statements that dealt with other dangerous chemicals. President John F. Kennedy, a regular reader of *The New Yorker*, apparently was familiar with the *Silent Spring* articles that summer. At a White House press conference on August 29, a reporter raised the issue.

"Mr. President," the reporter began, "there appears to be a growing concern among scientists as to the possibility of dangerous long-range side effects from the use of DDT and other pesticides. Have you considered asking the Department of Agriculture or the Public Health Service to take a closer look at this?"

"Yes, and I know they already are," the President replied. "I think particularly, of course, since Miss Carson's book, but they are examining the matter."

Because of the President's concern, the Life Sciences Panel of the President's Science Advisory Committee took over in earnest the study of pesticide use. Its report the following year became one of the milestones in the search for a satisfactory pesticide policy.

These and other reports filtered in to Rachel Carson, now resting at her summer place on the Maine coast. Biologists in the South wrote her to report meetings at local Agricultural Extension Service offices, ostensibly to organize pesticide control committees; in reality, the meetings turned into gabfests between agricultural workers and the representatives of chemical companies about the best means to counteract the message of *Silent Spring*. And from the Netherlands there arrived a letter from Dr. Briejèr. He reported that, even before he had seen *The New Yorker* articles, he had been asked about them by both his country's Minister of Agriculture and its Minister for Foreign Affairs.

"Commercial interests are strong," Briejèr wrote to her. "The use of herbicides is increasing and many complaints about damage are coming in. I am afraid many scientists in the field of plant-protection are on the wrong side."

For the time being, much of the adverse comment about *Silent Spring* was muted. The storm had not yet broken over Rachel Carson's head, as the chemical industry bided its time, waiting for the book's official publication when it could mount its counterattack under the guise of book reviews in both popular and scientific periodicals. Most of the comments now were congratulatory. A physician, writing to *The New Yorker*, remarked that Rachel Carson should feel deep satisfaction if her book received the attention it deserved.

"She will have accomplished as real a service as any physician who devoted a lifetime to patients," he wrote, "and she will have reached a 'practice' encompassing everyone!"

The gratification Rachel Carson experienced from this attention aroused by her book was tempered by dismay that she was now a "controversial figure."

> How dreary to be somebody!
> How public, like a frog
> To tell your name the livelong day
> To an admiring bog!

She shared this horror with Emily Dickinson. West Southport, where she had once felt secure, now seemed to her center stage. On August 1 she wrote to her friend Shirley Briggs in Washington: "So far I have not had to resort to disguises but I am about to have the telephone changed to an unlisted category, and even with that precaution I think I am about to be invaded by the press."

Her book was a major success before it arrived in the bookstores. It had been bought by the Book-of-the-Month Club (with Supreme Court Justice Douglas contributing an article on the book for the club's newsletter). Its advance sale had reached 40,000 copies on September 27, which was publication day. And then the storm broke.

*

What was this book which created such an uproar? *Silent Spring* is, essentially, an *ecological* book. Almost everything that had been said about chemical pesticides before this time had been phrased in *economic* terms: we need more and better pesticides to grow bigger and better crops to make agriculture more profitable and more convenient for the farmer. (That pesticides had proved a bonanza for the chemical manufacturers was the unstated element in this formula.) The ultimate statement, of course, was that any restrictions on the massive use of these "economic poisons" might deprive a struggling world of fibers to clothe, and food to feed, its needy millions.

The case for the unrestricted use of pesticides had been thoroughly aired. Rachel Carson approached the subject from a different direction — from the breadth of her experience in the biological sciences and the depth of her sympathy for all living things. In an address prepared for the Women's National Book Association after *Silent Spring* was published, she went to the heart of the matter:

"In each of my books I have tried to say that all the life of the planet is inter-related, that each species has its own ties to others, and that all are related to the earth. This is the theme of *The Sea Around Us* and the other sea books, and it is also the message of *Silent Spring*."

Growing more specific, she added:

"We have already gone very far in our abuse of this planet. Some awareness of this problem has been in the air but the ideas had to be crystallized, the facts had to be brought together in *one place*. If I had not written the book, I am sure the ideas would have found another outlet. But knowing the facts as I did, I could not rest until I had brought them to public attention."

In style and content, then, Rachel Carson designed *Silent Spring* to shock the public into action against the misuse of chemical pesticides. She described the poisons, pointed out the failure to grasp biological principles that allowed us to direct broadsides of these poisons against the environment, and detailed the resulting fiascos and disasters. (Her book included fifty-five pages of notes on her source materials.)

In her concluding chapters, Rachel Carson called for intensive research into the effects of these poisons (particularly their *long-term* effects) on all forms of life, including man. To restrict this massive infusion of chemicals into the environment, she wrote, will require the development and application of alternate methods of pest control. She called for sounder cultural methods to control insects and other crop pests, and the more selective use of weed killers. She was especially interested in biological controls — the attack on insect pests by artificially introduced diseases, predators, and sterilizers that attack the target species.

It will be important, as we rake over the abuse hurled at *Silent Spring*, to remember that Rachel Carson did *not* call for the abandonment of all chemical pesticides. She directed her charges chiefly at the long-lasting chlorinated hydrocarbon insecticides (like DDT) whose movement through the environment cannot be contained and whose residues, being fat soluble, are stored in animal tissues and recycled through food chains. For the other, shorter-lived pesticides, she counseled sanity and restraint in their use. Near the end of Chapter 2 in *Silent Spring* (page 11), the reader will find these words:

> It is not my contention that chemical insecticides must never be used. I do contend that we have put poisonous and biologically potent chemicals indiscriminately into the hands of persons largely or wholly ignorant of their potentials for harm . . . I contend, furthermore, that we have allowed these chemicals to be used with little or no advance investigation of their effect on

soil, water, wildlife, and man himself. Future generations are unlikely to condone our lack of prudent concern for the integrity of the natural world that supports all life.

The quality of the subsequent attacks on *Silent Spring* indicates that their authors either did not read, or did not understand, those words.

The chemical and agricultural industries saw *Silent Spring* not as a scientific challenge, but as a public relations problem. Their champions in the scientific world (many of them, in reality, were paid consultants to the industries) attacked the book on much the same grounds that, a century before, Louis Agassiz had challenged Darwin's *Origin of Species:* "A scientific mistake, untrue in its facts, unscientific in its method, and mischievous in its tendencies."

Many books of genuine quality might have been destroyed by the fury of an assault launched by powerful enemies. *Silent Spring* was able to survive the onslaught and take its place as an American classic partly because of the ineptness of the attacks on it by the agricultural-chemical clique, and partly by the skill of the prominent scientists who spoke out in its favor. The arguments of Rachel Carson's critics were characterized chiefly by the very "emotionalism" of which they had accused her, as well as by a reluctance to meet the issues. George Orwell, in an essay written in 1946, described just what we later observed in the *Silent Spring* controversy, for the quality of the various arguments somehow got reflected in the quality of the combatants' prose. Writing about the English language, Orwell remarked: "It becomes ugly and inaccurate because our thoughts are foolish, but the slovenliness of our language makes it easier to have foolish thoughts." If the reader doubts the objectivity of the assertions made above, a closer look at the uproar may help to convince him.

Some of the criticism aimed at *Silent Spring* makes amusing reading. F. A. Soraci, director of the New Jersey Department of Agriculture, had this to say in the *Conservation News* at the time of the book's publication: "In any large scale pest control program we are immediately confronted with the objection of a vociferous, misinformed group of nature-balancing, organic-gardening, bird-loving, unreasonable citizenry that has not been convinced of the important place of agricultural chemicals in our economy."

James Westman, chairman of the Department of Wildlife Conservation at Rutgers, preferred to base his attack on a ten-year-old article by William M. Foss in the *New York Conservationist,* from which he quoted the following: "As almost everyone knows, DDT — developed in 1874 but first used as an insecticide during the last war — is strong medicine, potentially dangerous to all forms of life. So is alcohol. But since we have become familiar with its properties, we have had no fear of using alcohol to suit our needs and purposes."

To which Westman added, "Amen." This sort of comment makes Rachel Carson look very good today.

Ironically, many of the attacks on Rachel Carson were prefaced by a bow to her "graceful writing." It was with this sort of gallantry that P. Rothberg, president of the Montrose Chemical Corporation of California (a manufacturer of DDT) introduced his remarks on *Silent Spring.* He went on to say that Rachel Carson wrote not "as a scientist but rather as a fanatic defender of the cult of the balance of nature." And William B. Bean, M.D., writing in *Archives of Internal Medicine,* went even further by saying he was sympathetic to Rachel Carson's cause. He added, however, that *Silent Spring,* "as science, is so much hogwash . . . I was made curious again and again by her disregard of the rubrics of evidence, of a nice regard for scientific validity, or of any feeling that what she presented should be unbiased."

But the only sentence in *Silent Spring* that he quotes to show her disregard of evidence is this one from Chapter 3 (page 13): "For the first time in the history of the world, every human being is now subjected to contact with dangerous chemicals from the moment of conception until death." Dr. Bean found this "an astonishing statement." Even more astonishing, however, were his own circumscribed views about the mobility of DDT residues. They have been discovered in remote regions of the world where spray planes have never intruded; they have been discovered even in mother's milk.

Many prominent scientists understandably were puzzled by the nature of the attacks on *Silent Spring*. Robert Cushman Murphy, the famous ornithologist from the American Museum of Natural History, recalled that he had taken part in a radio discussion of the book with Thomas H. Jukes of American Cyanamid Company. Jukes had called *Silent Spring* a "hoax." Then, in an article (written for a scholarly audience) in the *American Scientist* shortly afterward, Jukes had been effusive in his praise of chemical pesticides, yet did not criticize *Silent Spring* directly; he confined his attack to some rather dubious remarks that had been made about pesticides in an article in *The Police Gazette*.

The most widely distributed anti-Carson article that appeared in any scientific magazine was written by Dr. William J. Darby, a nutritionist at the Vanderbilt University School of Medicine, for *Chemical and Engineering News*. "Her ignorance or bias on some of the considerations throws doubt on her competence to judge policy," Dr. Darby said of Rachel Carson. "For example, she indicates that it is neither wise nor responsible to use pesticides in the control of insect-borne diseases."

Dr. Darby, then, like most of the book's critics, made a great show of refuting statements that Rachel Carson had never made. What she had questioned were the overall methods used in the war against insect-borne diseases; she did not rule out the use of pesticides in all cases.

The chemical industry presented an almost united front against what it considered the menace of Rachel Carson. There were allegations made at the time that certain chemical companies threatened to withdraw their advertising from gardening magazines and newspaper supplements that gave favorable mention to *Silent Spring*. In November, 1962, the Manufacturing Chemists Association began mailing monthly feature stories to news media, stressing the "positive side" of chemical use. Similar material was mailed to about 100,000 individuals. The National Agricultural Chemicals Association doubled its public relations budget. It distributed thousands of copies of reviews that were critical of *Silent Spring*.

This was the gist of the message: "A serious threat to the continued supply of wholesome, nutritious food, and its availability at present-day low prices is manifested in the fear complex building up as a result of recent unfounded, sensational publicity with respect to agricultural chemicals." In the face of even the mildest criticism, the chemical industry has resorted to this theme over and over in the intervening years.

The target of a considerable portion of this material (including "fact kits" which presented the answers wrapped up in the questions) was the medical profession. In the light of this unceasing barrage, it was little wonder that in November the *AMA News* suggested to interested doctors that they contact the chemical trade associations for material to help them answer their patients' questions about pesticides.

Rachel Carson, who was shocked to only a very small degree by the response her book had evoked, nevertheless felt compelled to speak out about this sort of thing. "I can't believe that the AMA seriously believes that an industry with $300 million a year in pesticide sales at stake is an objective source of data on health hazards," she told a reporter.

Meanwhile, an attack began to take shape from what at first appeared to be another quarter. Its source in this case was an organization called The Nutrition Foundation. This organization had been incorporated in 1941 to support fundamental research and education in the science of nutrition. As part of its educational activities early in 1963 it put together a "Fact Kit" on the subject of *Silent Spring.* The kit consisted of a defense of chemical pesticides prepared by the New York State College of Agriculture, and several book reviews that were critical of *Silent Spring.* (One of them was written by I. L. Baldwin, who was the chairman of the National Academy of Sciences-National Research Council's Committee on Pest Control and Wildlife Relationships.) It was accompanied by a letter, written by C. G. King, the president of the Foundation, which stressed the "independence" of those who attacked Rachel Carson's book, and described the book itself as "distorted."

"The problem is magnified," King said, "in that publicists and the author's adherents among the food faddists, health quacks, and special interest groups are promoting her book as if it were scientifically irreproachable and written by a scientist."

A man who could write that sentence obviously should not throw around the word "distorted." King was particularly disturbed because Rachel Carson did not share his admiration for the motives of some leaders in the chemical industry. In his letter, King said he had "known" many of these men, but he could have gone a little further and pointed out their connection with The Nutrition Foundation. The Foundation's membership consisted of fifty-four companies in the food, chemical, and allied industries, many of which depended on chemical pesticides for the cheap production of their raw materials. The presidents of many of these companies served on the Foundation's Board of Trustees. In the other direction, King might have pointed out that the Foundation funneled money, in the form of research grants, from these companies to the nutrition departments of many famous universities. Coincidentally, some of the fiercest attacks on *Silent Spring* came from university nutritionists. As one impartial nutritionist observed: "Where the shot hits, the fur flies."

The Foundation's "Fact Kit" was widely distributed. Included on the mailing lists were university people in many fields, researchers and other staff members of state agricultural experiment stations, the membership of the American Public Health Association, leaders of women's organizations, librarians, visiting nurses, and state, county and municipal officials. The kit was received by tree wardens in Connecticut, by the mayor of a small New Jersey borough, and even by a local Audubon Society in Pennsylvania. Monsanto itself did not operate more efficiently.

Industry supplied the personal touch, too. One of the most active figures on the lecture tour in the months following *Silent Spring*'s publication was Robert H. White-Stevens, assistant director of research and development in the Agricultural Division of the American Cyanamid Company. He made twenty-eight speeches before the end of 1962, extolling the virtues of pesticides and charging that *Silent Spring* was "littered with crass assumptions and gross misinterpretations."

White-Stevens was one of the first to counter the evidence of bird mortality in pesticide programs by a resort to the results of the annual Christmas Bird Census sponsored by the National Audubon Society. This dubious ploy, used many times since despite the Society's warning that it is invalid, points to the census to "prove" that the bird population of this country has actually increased since the introduction of chemical pesticides. It is ironical that those who question Rachel Carson's scientific procedures should equate with science an outing which even enthusiastic birdwatchers refer to as "bird golf."

The number of observers participating in the census, the purpose of which is to get a good score, has risen tremendously in recent years; naturally, the number of birds listed also has risen tremendously. The spectacular leap in the robin count for one year, often singled out for their purposes by industry spokesmen, was accounted for almost entirely by the discovery of a single huge gathering of these birds at a Nashville roost in the fall of 1960. In truth, the only significant increases noted recently among bird populations are among the blackbird-starling group. This increase happens to be due chiefly to such agricultural practices as land drainage and the development of large feedlots.

Rachel Carson allowed most of the attacks to pass over her head. To a friend she wrote: "Of course it is always tempting to answer these things but in the long run I believe it is better to let the course of events provide the answers." One incident, however, upset her because the criticism came from a scientist she respected, and she felt that it was caused by a misunderstanding.

As she had completed each chapter in her book, Rachel Carson sent copies of it for confirmation of various facts to experts in the particular field with which that chapter dealt. Among the scientists to whom she sent a copy of her chapter on the development of resistance to chemical pesticides by various insects was A. W. A. Brown, a zoologist at the University of Western Ontario. Along with the chapter, which she mailed to him on April 14, 1962, was a résumé of what her book would say. On April 23, Brown returned the chapter to her, telling her that it was well done and certainly brought out the point that she wished to make. He made a few suggestions for changes. He was most concerned about the title of the chapter which she had called "The Rumblings of an Avalanche"; Brown felt this to be an inept simile for the process of insect resistance, suggesting that it was closer to the movement of a glacier than to that of an avalanche.

When *Silent Spring* was published, however, Brown apparently was horrified. His displeasure spilled over into the Toronto newspapers, where he claimed in an interview that his suggested changes were not made. He felt that Rachel Carson, by quoting him, had put him in a bad light with his colleagues. He was quoted in the Toronto press as saying that "individuals concerned with wildlife work, having a vested interest in opposing pesticides," had pressured Rachel Carson into taking an extreme position. This was profitable for Miss Carson, he suggested, but hard on some "competent public servants" who had to pay for it all.

The report in the newspaper prompted at least one prominent

Canadian official to rise to Rachel Carson's defense. C. H. D. Clarke, Chief of the Fish and Wildlife Branch of the Department of Lands and Forests, wrote to Brown:

> I suggest to you that every intelligent and sensitive person in the world has a vested interest in wildlife; and that we have a responsibility for the interest of those not yet born . . . If Miss Carson's book has sold well it is only because so many people have a vested interest in wildlife and are concerned with their heritage . . . I am at a loss also to know what public servants have had to pay, unless it is more attention to what they are doing.
>
> . . . In my view, the essence of ecology is the consideration of time, and all the wildlife people have ever done is point out to the chemists that their particular assassination has not trammeled up the consequences, and nobody knows, though we have a right to know, what its surcease will be.

This was one instance in which Rachel Carson did not stand aside and allow her friends to carry the burden of her defense. She too felt the facts were being misrepresented, and not by her. She went over her correspondence with Brown and determined that, aside from the matter of the chapter title, she had made all but one of the changes he had suggested, and that one was "a minor point and not a matter of factual accuracy."

> I am sorry to hear that you feel as you do [she wrote to Brown], but in all frankness I must say that the time to voice those objections was a year ago when the manuscript was under consideration.
>
> Quotations and views attributed to you were all drawn from published papers, to which your colleagues presumably had access, so I do not know why they are surprised to read them now. If you had in the meantime changed your mind on some of these matters, you could have requested that I delete any such statements when you reviewed the manuscript and I would of course have respected your wishes. I find no such request in the correspondence, however.

> Yes, I recall that you disagreed with the concept of an "avalanche" but I must point out two facts: first, that titles are clearly the prerogative of the author, and further that the phrase was specifically attributed to Elton [Charles Elton, the British ecologist], not to you.

Silent Spring undoubtedly upset a great many people. Part of the attack on it, of course, was launched because some of these people believed that any restriction, however slight, in the profligate use of pesticides would inevitably cost them a dollar. But there were others who honestly felt affronted by certain aspects of the book. Until *Silent Spring*'s publication, for instance, almost every public statement on pesticides (as we have seen in the case of the first two National Academy of Sciences reports in 1962) had been overbalanced in favor of the chemical control of pests. They had been simply litanies in praise of pesticides. When Rachel Carson neglected to give equal time to the achievements of pesticides in her book, some of her colleagues accused her of being "unscientific."

But Rachel Carson saw no reason to pay homage to pesticides in this particular book. As Clarence Cottam asked in the Sierra Club *Bulletin*, "Hasn't this side of the problem already been overemphasized by a multi-billion dollar industry employing the most experienced salesmen and lobbyists available?"

It was Rachel Carson's intention to tell the side of the story which had not yet been told. Like any good prosecutor or lawyer, she looked into the *merits* of the case before she took it on. In doing so, she built, on a basis of solid fact, the first outstanding work on ecology that had yet reached a wide popular audience. This accomplishment was perhaps best described by Paul B. Sears, Professor Emeritus of Conservation at Yale, who wrote of *Silent Spring*: "The result, over and above her usual clarity of structure and presentation, is a brief of which any attorney might well be proud. If anything can convince the court of public opinion, this should do so. She has also produced an important contribution to ecology."

Yet this aspect of *Silent Spring* was overlooked completely by those who were only too eager to find soft spots in her argument. (If the reader believes that a "true" scientist never comes under attack from fellow scientists, he can disabuse himself of this notion by thumbing through the pages of any scientific publication; it is likely to bristle with internecine warfare.) In the belief that Rachel Carson had taken leave of her scientific senses, her critics flailed out at her. The result has been described by Cottam:

"Miss Carson has been referred to slightingly as a priestess of nature, a bird-, cat-, or fish-lover, and a devotee of some mystical cult having to do with the laws of the Universe to which critics obviously consider themselves immune."

Other scientists, though willing to accept the documented bulk of *Silent Spring*, were intellectually capsized by the book's opening chapter. This brief opening (679 words) was called "A Fable for Tomorrow." In it Rachel Carson created a town and destroyed it under a rain of pesticides. (Originally she had even given her town a name, "Green Meadows," but was dissuaded by her editor, who felt that the name "suggests not an old town but a real estate development.")

"I know of no community that has experienced all the misfortunes I describe," she wrote. "Yet every one of these disasters has actually happened somewhere, and many real communities have already suffered a substantial number of them."

It was this brief opening which prompted some of its critics to relegate *Silent Spring* to the science fiction shelf.

"It just 'turned off' many scientists," Roland C. Clement of the National Audubon Society has said. "The chapter is an allegory. But an allegory is not a prediction, which is what the literal-minded readers, with no background in literature, confused it with."

This opening chapter provided the basis for a long parody in *Monsanto Magazine*, the house organ of the Monsanto Chemical Company. The parody, entitled "The Desolate Year," depicted the horrors visited upon a world in want of chemical pesticides. Monsanto's public relations men began working on the parody after the excerpts of *Silent Spring* appeared in *The New Yorker*. Because Rachel Carson's book was scheduled for publication before the next issue of *Monsanto Magazine*, the public relations men cleared the parody with the company's officials and distributed 5,000 sets of galley sheets to newspaper editors and book reviewers all over the country.

"This was, for us, an opportunity to wield our public relations power," a Monsanto man said.

Later, LaMont C. Cole, Professor of Ecology at Cornell, was to speak of Monsanto's grisly vision of a world without pesticides in his review of *Silent Spring* in the *Scientific American*: "It is easy to become persuaded that years like those just before World War II could not possibly have occurred: no chlorinated hydrocarbons, no organic phosphates, no payments to farmers to reduce production and still crop surpluses!"

Industry's contention that the scientific world rejected *Silent Spring* was belied by the number of outstanding scientists who declared their admiration for the book. This movement occurred despite the traditional view of the scientist as a man who will not stick his neck out if the cause does not immediately concern him; evidently, many scientists believed that environmental contamination was of immediate concern to them. Robert L. Rudd wrote to Rachel Carson to tell her of the reaction to *Silent Spring* at the University of California.

"I don't have to tell you that *Silent Spring* brought everyone out of the woodwork," Rudd wrote. "The range of opinion in the UC system is as extreme as anywhere else. You'd be surprised, nonetheless, how very much sympathy the points of view we have both held have with some of our specialists in the [Agricultural] Experiment Station and among the scientific disciplines."

Across the country, Loren Eiseley, the celebrated anthropologist at the University of Pennsylvania, added his voice to the support for Rachel Carson. Reviewing her book for the *Saturday Review,* he wrote: "If her present book does not possess the beauty of *The Sea Around Us,* it is because she has courageously chosen, at the height of her powers, to educate us upon a sad, an unpleasant, an un-beautiful topic, and one of our own making. *Silent Spring* should be read by every American who does not want it to be the epitaph of a world not very far beyond us in time."

One of the most balanced reviews of the book was written by LaMont C. Cole, referred to above, in the *Scientific American.* His exhaustive examination of the book allowed him to destroy the heavy-handed attacks on it. Of *Time* magazine's review, he said: "Without citing evidence, (it) proclaims the merits of pesticides in a statement with which, in my opinion, no responsible scientist would want to associate himself."

Whereas the industry-inspired reviews of *Silent Spring* had simply howled indignantly about "distortion," or attacked Rachel Carson for statements she had never made, Cole pinpointed what he believed to be errors in her work. He questioned her statement on page 16 that rotenone and pyrethrum are "simpler inorganic insecticides of prewar days." Pyrethrum, Cole pointed out, is so complicated a substance that "it has frustrated the analysis of the ablest organic chemists," and he suggested that its complicated nature may be "also responsible for frustrating the abilities of insects to develop resistance."

Elsewhere, Cole complained that Rachel Carson had not mentioned the fact the "bees are probably less threatened by modern insecticides than they were by the older arsenicals." He also questioned her statement that resistance among insects causes us to seek ever more potent poisons. The insects that have become resistant to DDT prove less adaptable in other respects, and fall victim to weaker, unrelated pesticides. "I do not for a moment believe," Cole wrote, "that the chemicals are producing super insects." *

The attempt to undermine *Silent Spring* by questioning the facts on which Rachel Carson based her case was doomed to fail. "Errors of fact are so infrequent, trivial and irrelevant to the main theme that it would be ungallant to dwell on them," Cole said. Of Rachel Carson's overall treatment of the use of chemical pesticides, he concluded: "She does not convey an appreciation of the really great difficulties of the problem . . . But what I interpret as bias and oversimplification may be just what it takes to write a best-seller . . . If the message of *Silent Spring* is widely enough read and discussed, it may help us toward a much needed reappraisal of current policies and practices."

Silent Spring was an enormous undertaking, as any work is that tries to bring together many disciplines to create a workable synthesis. Rachel Carson saw what most "pest control experts" had not seen — that the specialized view cannot solve the many

* Points which other scientists have raised about inaccuracies in *Silent Spring* include the statement on page 15 that arsenic in chimney soot is the cause of cancer; many, though not all, scientists believe it is the tars in soot that are the chief carcinogens. The role of arsenic, however, as a potent human carcinogen is firmly established. According to Dr. W. C. Hueper of the National Cancer Institute, "The recent severe epidemic of arsenic cancers among German vintners which developed as the result of their use of arsenical pesticides between 1920 and 1940 and which has given rise to the appearance of often multiple and multicentric cancers of the skin, lung, liver, esophagus, larynx, nasal sinuses, etc., should be adequate and ample to convince even the professional sceptics on this point."

Rachel Carson seems to have overstated the threat to robins during the frenetic assault on the elm bark beetle. DDT, used in the control of Dutch elm disease, has taken an enormous toll of robins, but ornithologists see no immediate possibility that the species will plunge "into the night of extinction" (page 86). Certain other species, as we shall see, certainly face extinction.

Perhaps her statement on page 238 that insect pathogens biologically "do not belong to the type of organisms that cause disease in higher animals or in plants" is too general. According to Edward A. Steinhaus of the Center for Pathobiology at the University of California at Irvine, "there may be rare exceptions in which insect pathogens are found to be pathogenic for vertebrate animals; the reported infection of a horse by the fungus *Entomophthora coronata* being a case in point."

problems posed by the large-scale use of pesticides. Indeed, such a limited view contributes to the problem. The variety of forms in nature baffles and blinds even scientists, just as the wealth of vegetation in the deep woods shuts off a man's view of all the surrounding forms except those closest to him. It is natural for the specialist to resent the overview. For her temerity, Rachel Carson bore the burden of a great deal of this sort of resentment. The University of California's Robert L. Rudd answered these critics of Rachel Carson when he reviewed *Silent Spring* in *Pacific Discovery* (Nov.-Dec. 1962), the publication of the California Academy of Sciences:

"Are they correct? I should say, 'Yes, in part,' if what is expected is an ultimate knowledge of every aspect of the problem. However, no reviewer, including her critics, has that total knowledge today . . . In my opinion, she is eminently qualified to present the facts, synthesis and argument she has in *Silent Spring*. I leave it to her critics to do as well."

And we will leave it to Robert Rudd, whose own exhaustive book, *Pesticides and the Living Landscape* (based in large part on his own years of experience in the field and laboratory), appeared shortly afterward, to define the essence of Rachel Carson's book:

"*Silent Spring* is biological warning, social commentary and moral reminder. Insistently, she calls upon technological man to pause and take stock."

5. The Public Controversy

BECAUSE OF RACHEL CARSON'S STATURE, the pesticide controversy was no longer confined to the technical journals. It had broken away into public debate. *Silent Spring* leaped onto the best-seller lists almost immediately after publication. By December more than 100,000 copies had been sold in the bookstores. Besides the Book-of-the-Month-Club edition, the Consumer's Union made available a special paperbound edition of its own for its subscribers.

To her discomfort, but certainly not to her surprise, this retiring woman had become the center of national controversy. Editorials in the country's leading newspapers generally were favorable to *Silent Spring*. Conversely, the articles and reviews in national magazines were not. An article in the Science section of *Time* set the tone. "Miss Carson has taken up her pen in alarm and anger, putting literary skill second to the task of frightening and arousing readers," the article began. And it concluded: "Many scientists sympathize with Miss Carson's love of wildlife, and even with her mystical attachment to the balance of nature. But they fear that her emotional and inaccurate outburst in *Silent Spring* may do harm by alarming the nontechnical public, while doing no good for the things that she loves."

If later events have proved this early *Time* statement wrong, its editors were no farther off the mark than those of most of the other major periodicals. In this case, God definitely was not on the side of "the big battalions." It was the responsible elements of the scientific community that rushed to Rachel Carson's defense, and helped to set the record straight. Roland C. Clement of the National Audubon Society characterized the nature of the powerful, public-relations-based attack on *Silent Spring*. Speaking of those magazine articles that champion pesticides so "that no semblance of objectivity remains," Clement said:

> One of these is the hoax of "bigger and healthier wildlife populations, not in spite of pesticides but in many cases because of them," that *Sports Illustrated* took hook-line-and-sinker from one of its lady staff writers. One cannot avoid wondering whether some clever PR man planted this with the authoress so that the fair sex might seem to help "right the record" against Rachel Carson.*
>
> The next item, again, tries to pass off Mr. Edwin Diamond as an early collaborator of Miss Carson's, and one who left her in disillusionment. Even though the *Saturday Evening Post* made this claim for Mr. Diamond the facts seem to be, according to Houghton Mifflin Company which employed Mr. Diamond and published *Silent Spring*, that this man was merely assigned to help "research" certain pesticide problems and never worked *with* Miss Carson at all. It was Houghton Mifflin that parted company with Mr. Diamond.
>
> Next is *The Reader's Digest* account of "The Great Pesticide Controversy" by Strohm and Ganschow, which turns out, on careful reading, to be another assurance that everything is all right by authors who don't know how to ask questions.

* By 1968 both *Time* and *Sports Illustrated* viewed matters in a different light. In an essay in their May 10, 1968, issue, the editors of *Time* wrote: "DDT is almost certainly to blame for the alarming decrease in New England's once flourishing peregrine falcons, northern red-shouldered hawks and black-crowned night herons . . . One of the prime goals in attacking pollution ought to be a vast shrinkage of the human impact on other creatures. The war on insects, for example, might actually go a lot better without chemical pesticides that kill the pests' natural enemies, such as birds. One of the best strategies is to nurture the enemies so they can attack the pests; more insect-resistant crops can also be developed." Almost simultaneously, in several hard-hitting articles, *Sports Illustrated* repudiated the notion that DDT is good for wildlife.

Rachel Carson cherished the support given to *Silent Spring* by outstanding biologists like Clement, Cottam, and Rudd. Likewise, she cherished the support of such internationally known scientists as LaMont C. Cole, Loren Eiseley, Julian Huxley, Charles Elton, and Hermann J. Muller. Considerably less exhilarating to this exacting woman was the applause from another quarter. Rachel Carson always tried to keep at arm's length those food faddists, health quacks and other cultists with which her severest critics tried to link her. Though often alarmed by the blunders of techno-science, she was a scientist herself and respected the critical approach to the world. Inevitably, she was made uncomfortable by those semi-mystical groups which detect a murderous intent behind any use of modern chemicals.

"They really dismayed her," her friend Shirley Briggs has said. "Rachel was mortified when nutty little publications praised her book — generally for all the wrong reasons."

While it was this disinclination to become involved with far-out groups that prompted her to decline some invitations to accept awards and speak at banquets, her failing health was far more often the cause. Invitations poured in on her. One of the few affairs she attended that winter (and this only because of persistent wheedling by close friends) was the Women's National Press Club dinner in Washington on December 5, 1962. On this occasion she made one of her infrequent replies to the critics of *Silent Spring*.

"(One) reviewer was offended because I made the statement that it is customary for pesticide manufacturers to support research in chemicals in the universities . . . I can scarcely believe the reviewer is unaware of it, because his own university is among those receiving such grants."

She went on to quote from research contributions to the *Journal of Economic Entomology*, in which the authors acknowledged financial support from such prominent chemical firms as Shell, Velsicol, and Monsanto.

It is possible, of course, that had her health been better Rachel

Carson might have struck more frequent blows in defense of her book. Certainly she was proud of the achievement of *Silent Spring* and resented any attempt to diminish it. One such example of the intense pride which she mingled with a jealous regard for her privacy occurred that winter. A state conservation officer mailed her the proof of a short profile he had written about her, intending to use it as background material during a symposium on pesticides. He asked her to make any corrections she thought necessary, and return the corrected proof to him. Rachel Carson made several deletions of items she believed too personal: references to her wearing "a small hat," to her habit of keeping Thoreau's *Journal* "near her bedlamp," and to the fact that she lived in Maryland "with her cat." Near the end of the profile was a reference to *Silent Spring* and a statement that "only a few minor inaccuracies have been pointed out in her recent book." She underlined the word "inaccuracies," and pencilled heavily in the margin the terse query, "WHAT?"

But by this time (and despite the continuing assault on it) *Silent Spring* had made its enormous impact on public life. Candidates for public office wrote to her, asking that she suggest appropriate stands for them to take on the pesticide issue. By the end of 1962, there had been over forty bills introduced in the various state legislatures to regulate pesticide use. Kansas and Iowa pushed through laws which required that professional applicators of pesticides be licensed.

"There is already a move afoot to have a legislative investigation of pesticides in Connecticut," a biologist in that state told her in a letter. "If this comes off, it will be solely because you 'softened up' the public for this needed reform."

Kenneth M. Birkhead, Assistant to the Secretary of Agriculture, assured a concerned U.S. Senate that the USDA had undertaken a series of steps to find "better methods to control pests." USDA's statement admitted that "Miss Carson presents a lucid description of the real and potential dangers of misusing chemical pesticides." Although it included the usual bow to the new chemicals ("the burst of productivity over the past decade on U.S. farms parallels the increasing use of chemical pesticides"), the statement also emphasized USDA's expanded research to develop biological controls and other methods "of controlling insects without the use of chemicals that leave harmful residues."

Often such statements (before and *after* the publication of *Silent Spring*) have been merely rhetorical crumbs to divert critics of the current policies and practices. But buried within the USDA statement was word of a tangible step toward a saner pesticide policy. This was the announcement that the department would request Congress to amend the Federal Insecticide, Fungicide and Rodenticide Act to abolish the so-called "protest registration." That such a loophole existed so recently in the federal control of pesticides remains a source of amazement to concerned conservationists even today. This provision of the Act dealt with pesticides, already on the market, that eventually were condemned by the government as excessively hazardous. The manufacturer of the offending product, notified of this by the government, nonetheless was able to keep it on the market simply by appealing USDA's decision. Only when the appeal finally had been disposed of, and the product's hazardous nature confirmed, was the manufacturer required to recall it from public sale. (The amendment that altered the "protest registration" loophole in 1964 was a milestone in the chronicle of pesticide legislation; the code of responsibility was turned inside out, so that instead of the government being obliged to prove *hazard*, the manufacturer now must prove *safety*.)

Silent Spring and its message permeated various corners of American life. Somebody wrote and recorded a song called "Silent Spring," which was unauthorized, of course, by Rachel Carson. In her nationally syndicated column, Sylvia Porter reported that dozens of "safe insecticides" were hastily concocted by gyp artists in the wake of the book's publication. Miracle products, phony experts, and ultrasonic rodent control devices proliferated. Miss Porter described the marketing of a "nonpoisonous" fertilizer, which turned out under analysis to be simply a potion of green dye and ordinary fertilizer "which streaked and washed off with the first rain." At the same time, there was an increase in the use of such legitimate but "old-fashioned" botanical pesticides as pyrethrum and rotenone.

Edwin Way Teale wrote to Rachel Carson at the end of 1962, passing along the latest gossip: "Here are two things to add to the growing legend. I heard at the last NY Entomological Society meeting that Rachel Carson was a kind of Joan of Arc who had lost her job at the Fish and Wildlife Service because she wouldn't give up the idea of writing *Silent Spring*. Also heard that a woman in Bucks County, Pa., read the book and got so mad at a neighbor who sprayed and some of the spray blew over on her property that she quit going to a Quaker meeting because she would have to speak to him there."

And of course the interviewer or admirer who somehow got through to Rachel Carson herself invariably asked the question: "And what do *you* eat?"

To which Rachel Carson invariably replied: "Chlorinated hydrocarbons, just as everybody else does."

*

In the early spring of 1963, Connecticut's Senator Abraham Ribicoff wrote to an interested citizen: "Since January, I have given a great deal of thought to this subject, and I have read Miss Carson's book carefully. I have concluded that what is needed is a broad gauge Congressional review of all Federal programs related to problems of environmental hazards, including pesticide control, air and water pollution, radiation and others."

Ribicoff happened to be a member of the Senate Subcommittee on Reorganization, which has jurisdiction over the inter-agency coordination of various Federal programs. Accordingly, he suggested such a review to Senator Hubert H. Humphrey, the subcommittee chairman. Humphrey agreed that the review was needed and on April 4 announced that it would take place, naming Ribicoff as acting chairman of the subcommittee for the purpose.

While *Silent Spring* and the issues it raised were certified for debate on Capitol Hill, there seemed to be some doubts in industry about its fitness as a suitable subject for family viewing. Fred Friendly, the executive producer of the Columbia Broadcasting System's television documentary series, "CBS Reports," scheduled a program on the subject for April 3, 1963. It was called "The Silent Spring of Rachel Carson." The first indication that this might prove to be something out of the ordinary came to CBS with the arrival of what Friendly called "an unprecedented volume" of advance mail. More than 1,000 letters poured in on CBS to discuss a program that had yet to be shown. Most of the letters expressed the wish that CBS would not show the program, although there is no proof that the anti-Carson campaign was organized.

Then, shortly before the day on which the documentary was to be shown, three of the program's five sponsors withdrew. Standard Brands, Inc., the manufacturers of various food products, Lehn and Fink Products Company, the manufacturers of Lysol, and the Ralston Purina Company, the manufacturers of a number of products including animal feeds, found *Silent Spring* strong stuff. Standard Brands thought the subject "incompatible" with its line of products. A spokesman for the firm said Standard Brands had withdrawn once before, from a program about bull fighting which was concerned with "blood and gore." Lehn and Fink also cited "incompatibility" as the grounds for its temporary separation from "CBS Reports." Instead, it transferred its sponsorship for that evening to a show about a talking horse. Apparently the horse seldom discoursed on controversial subjects.

Two sponsors remained at airtime — the Kiwi Polish Company and the Brillo Manufacturing Company. Rachel Carson appeared on the program, as did Dr. Luther Terry, the United States Surgeon General; Orville L. Freeman, the Secretary of

Agriculture; George Larrick, the Commissioner of the Food and Drug Administration; John Buckley of the Interior Department; and Robert White-Stevens, American Cyanamid's all-purpose spokesman.

The program brought out the fact that America was loading its environment with pesticides without the faintest idea about where it all led. The next day, Abraham Ribicoff rose on the floor of the Senate to point out this to his colleagues. "As last night's CBS telecast clearly showed," Ribicoff said, "there is an appalling lack of information on the entire field of environmental hazards. We face serious questions, but we are woefully short of answers."

This notion (which Rachel Carson had developed in her book) was dramatically confirmed the following month. A special panel of President Kennedy's Science Advisory Committee had labored for eight months on its inquiry into pesticide use. Friends in government kept Rachel Carson informed about its deliberations. Afterward she was able to write to a friend that President Kennedy "often asked about the progress of the Committee and urged speed in getting out the report."

Rachel Carson was among the experts invited to discuss the pesticide problem with the committee. The committee also called in scientists from the Department of Agriculture, and asked the chemical industry to designate spokesmen to present its side of the story.

"The people from the Department of Agriculture were very much conditioned to emphasize the 'Armageddon' which faced the world if pesticides were restricted," recalls William H. Drury, Jr., Director of the Hatheway School of Conservation Education and a member of the committee. "If you differed with the use of pesticides it was an all-or-nothing decision. We could *not* get them to talk in terms of drawing back certain pesticides or reducing certain pesticide use. They insisted on talking about using pesticides or not using pesticides. The people from the chemical companies played it a lot cooler. They were talking about their safety techniques and how they had met the requirements of the Food and Drug Administration, and the financial problems of developing a pesticide."

There was no basic disagreement within the committee. Despite frequent attempts by officials of the Department of Agriculture and the Public Health Service to dilute it (in keeping with their desire for a free hand in conceiving and carrying out control programs) the committee moved always in the direction of a strong report.

To write a strong report required among the committee members a view that contradicted the traditional one of the pest control interests. The committee held that there were inherent hazards in the current use of chemical pesticides, even in their *approved* use. Further, they recognized that science was not equipped with the knowledge at that time to assess accurately the full extent of those hazards. In writing their final version of the report, the committee members apparently considered the *lack* of knowledge in some areas to be as significant as the accumulation of facts in other areas.

The report, then, of the President's Science Advisory Committee (PSAC) proved to be a critical examination of both the values and hazards of pesticides. Because of the poorly balanced nature of most of the previous government statements, the PSAC report seemed revolutionary to both the conservationists and the pest control interests. It echoed Rachel Carson's criticism of the federal government's "eradication" programs, the use of persistent pesticides, and the general lack of concern for human safety. Of special importance to those who had been calling for a sound pesticide policy were the committee's recommendations. Among its recommendations were these:

- The Department of Health, Education and Welfare should cooperate with other agencies in a continuing network to monitor the level of pesticide residues in air, water, soil, fish, wildlife, and man.
- The Department of the Interior should considerably in-

crease its research and evaluation of the toxic effects of pesticides on wild vertebrates and invertebrates.

• The Department of Agriculture should redouble its efforts to develop both more selective chemicals and alternate methods of pest control.

• All government agencies should, at the conclusion of each of their large-scale pest control programs, issue a report so that other scientists and the public will be able to see for themselves how the decisions were made, what was accomplished, and how the agencies justified their programs.

Perhaps the most far-reaching note the committee sounded was its famous "Recommendation Five," which read as follows:

> The accretion of residues in the environment should be controlled by orderly reduction in the use of persistent pesticides. As a first step, the various agencies of the Federal Government might restrict wide-scale use of persistent insecticides except for necessary control of disease vectors. The Federal agencies should exert their leadership to induce the states to take similar actions.
>
> Elimination of the use of persistent toxic pesticides should be the goal.

This recommendation, as we shall see, has remained the keystone of the subsequent drive by concerned officials and citizens to halt the use of DDT and its related chemicals. The PSAC Report closed with the less-than-dramatic, yet equally significant, recommendation:

> To enhance public awareness of pesticide benefits and hazards, it is recommended that the appropriate Federal departments and agencies initiate programs of public education describing the use and the toxic nature of pesticides. Public literature and the experiences of Panel members indicate that, until the publication of *Silent Spring* by Rachel Carson, people were generally unaware

of the toxicity of pesticides. The Government should present this information to the public in a way that will make it aware of the dangers while recognizing the value of pesticides.

The PSAC Report was issued on May 15, 1963. "RACHEL CARSON STANDS VINDICATED," ran a headline in *The Christian Science Monitor.* Eric Sevareid, the CBS news commentator, told his audience that "Miss Carson had two immediate aims. One was to alert the public; the second, to build a fire under the Government. She accomplished the first months ago. Tonight's report by the Presidential panel is *prima facie* evidence that she has also accomplished the second."

Technical publications (many of which had criticized *Silent Spring*) joined the popular press in applauding the PSAC Report. *Chemical and Engineering News* noted that "The committee's panel on the use of pesticides was composed of men of achievement in scientific and public affairs, whose positions imply recognition of their judgment and responsibility. The tone of their report reflects those qualities." And *Science*, the magazine of the American Association for the Advancement of Science which had published only a few months before I. L. Baldwin's highly critical review of *Silent Spring*, made this remarkable statement on its editorial page: "The long-awaited pesticide report of the President's Science Advisory Committee was issued last week, and though it is a temperate document, even in tone, and carefully balanced in its assessment of risks versus benefits, it adds up to a fairly thorough-going vindication of Rachel Carson's *Silent Spring* thesis."

Standing a little glumly on the sidelines was the National Agricultural Chemicals Association. Its spokesman told a *Wall Street Journal* reporter that the PSAC Report contained "observations and surmises which have never been confirmed by scientific investigation." He indicated that reasonable doubt and a scarcity of precise information were not acceptable as a basis for public policy. The trade association felt that the PSAC Report called for more caution than industry wanted, but seemed to be relieved that its criticism and recommendations were not more severe.

In industry's eyes, at any rate, the presumption of innocence until proof of guilt had been established remained applicable to economic poisons. The controversy, as well as the damage, continued.

6. The Widening Protest

In June, C. J. Briejèr, the plant specialist, wrote to Rachel Carson from the Netherlands. "The hurricane you unchained is now over us," he told her. "The chemical industry is furious and so are several chemical-minded scientists. You might understand this made me a bit nervous and therefore I was extremely pleased in receiving the Report of your President's Science Advisory Committee, a very strong weapon for repelling all attacks."

The events that had agitated American scientific circles during the previous year had not gone unnoticed abroad. *Silent Spring* had been translated into twelve languages. Its message fell on receptive ears. Writing, several years later, about the book's publication in England by Hamish Hamilton, Ltd., Julian Huxley recalled that "the situation in Britain at that time was as grave as in the United States. As a result of insecticides and herbicides, we are in the process of losing large numbers of our songbirds and birds of prey, our butterflies and bees, and our prized wild flowers. When I told my brother Aldous the facts, he said, 'We are destroying half the basis of English poetry.' "

While the debate over *Silent Spring* and its message flared in American legislative and academic halls, a similar debate was taking shape in the House of Lords in London. The same symptoms (if on a smaller scale) of the irresponsible broadcasting of chemicals had appeared in Great Britain. After the publication there of *Silent Spring*, public concern caused the issue of the use of toxic chemicals to be paraded in full debate in the House of Lords during the spring of 1963. By official count, Rachel Carson or her book were mentioned twenty-three times during the proceedings. Peter Scott, the English naturalist, recently spoke of the debate.

"I well remember watching from the public gallery the debate in the House of Lords and seeing two red spots below — the dust jackets of the English edition of *Silent Spring* beside the dispatch box on either side — one for reference by the Government spokesman, Lord Hailsham, the other by the Opposition spokesman, Lord Shackleton. Nothing has since occurred to modify my view that future generations will regard Rachel Carson as a great benefactor to the human race for the impact created by *Silent Spring*."

One of Rachel Carson's admirers in the House of Lords sent her a copy of the official report of the debate that spring. "You will notice," he told her, "that almost every speaker in the debate mentioned you and *Silent Spring*. This is something which has never happened before in my experience of Parliament and I hope you will understand it as proof that your book has made a strong impression in this country."

What was the background of *Silent Spring*'s success in Great Britain? The pesticide problem had not raced out of control there as it had in the United States. Monoculture is not practiced on so vast a scale; biting insects, especially in southern England, are not a major nuisance; and malaria disappeared from the island before the introduction of chemical pesticides. British scientists often expressed amazement at profligate American pesticide programs, such as those directed at the fire ant and the gypsy moth.

Nevertheless, the British scene contained sufficient ingredients for an environmental disaster. The British were as amply impressed as Americans by DDT's early triumphs. Their laws are as firm as America's in upholding private (*i.e.* business) privilege against the public welfare. Note these words of a distinguished committee, justifying its do-nothing stance, after investigating London's serious air pollution problem in the early

1920's: "We have never ceased to bear in mind that the interests of trade must be fully considered, and that the introduction of legislation which might prejudicially affect important industries is quite out of the question." This sort of weakness in the face of corporate power led inevitably to London's infamous "killer smog" in 1952.

A few years later the government's unconcern about the ultimate destination of the new chemical pesticides, then being widely broadcast across England's green and pleasant land, brought on another environmental crisis. Each spring, beginning about 1956, corpses of birds were seen everywhere in the landscape. The owners of large estates found them by the hundreds — chaffinches, greenfinches, linnets, and hedge sparrows, as well as pheasants and partridges. Witnesses told of seeing pigeons falling dead from the sky. In 1961, Lincolnshire, an agricultural region, reported over 10,000 dead birds, perhaps only a fraction of the full total there. Foxes and other mammals, feeding on the carcasses, died in their turn.

Shocked citizens prompted both scientific and parliamentary investigations. Fortunately, there were mechanisms already available for dealing with the crisis. Pesticides had been under suspicion in England for some time. In 1957, a plan which eventually became known as the "Pesticides Safety Precautions Scheme" was negotiated between government departments and the principal trade associations involved in the manufacture and application of chemical pesticides. Under the terms of the scheme, manufacturers and distributors agreed not to introduce new pesticides, nor to advocate new uses for existing ones, until the pesticides and their uses were approved by a committee of scientists. The plan was entirely voluntary.

As reports of deaths among birds and mammals became more common, the Nature Conservancy (the official department which is responsible for advising the British Government on matters of conservation) took a more lively interest in pesticides. In 1960 it created a Toxic Chemicals and Wildlife Division to study the effects of pesticides on wildlife. The division became based at the Monks Wood Experimental Station in Huntingdon. At nearly the same time a joint committee of the British Trust for Ornithology and the Royal Society for the Protection of Birds (two private conservation organizations) was established. The committee published reports on pesticide effects among birds, and began to alert the public to what one naturalist called "the biggest risk to wildlife and game that ever occurred in the country."

In 1961 the facts were revealed to the English public. Chlorinated hydrocarbon pesticides, especially dieldrin, were being used to coat cereal seeds before planting in order to protect them from the wheat bulb fly. Birds of many species either dug up the planted seeds or ate those they found lying on the surface of the fields. Spectacular mortality among both game and song birds was the inevitable result.

When similar damage has been uncovered in the United States, the agri-chemical interests usually have denied the existence of a problem, or have shrugged it off as being justified by cost/benefit considerations. Once the facts were clear in England, however, no segment of society shirked its responsibilities. Few individuals were deceived by the specious "birds or people" argument. There was, instead, an ecological awareness that the suffering and death of birds over large areas was an urgent warning to men. Just as pain is a signal that something has gone wrong in the body, this widespread destruction was interpreted as a signal that the environment had sickened.

The initial impact of *Silent Spring*, to which the British were especially sensitive, contributed to the solution of the problem there.

"Rachel Carson's book did much to stimulate general interest in the problem," says N. W. Moore of the Nature Conservancy. "And, doubtless, like the efforts of the voluntary ornithological bodies, directly and indirectly provided support for research."

In 1962, even before *Silent Spring* was published in England, a voluntary agreement went into effect among pesticide manufacturers, farmers, and government officials not to use the harmful chlorinated hydrocarbons as cereal seed dressings during spring sowing. These pesticides might be used during the fall sowing because at that time there are a great many natural food sources for birds. By late winter and early spring, however, the food sources have largely disappeared. It is then that the birds will resort to digging up seeds in promising areas.

Since *Silent Spring*'s publication there, England has paid close attention to the dispersal of its pesticides. Most other field uses of the offending chlorinated hydrocarbons (aldrin, dieldrin, and heptachlor) were banned voluntarily in 1966. In the same year a ban was imposed on the use of aldrin and dieldrin as sheep dips to combat certain flies whose maggots feed on the flesh of sheep.

Meanwhile the industry continues to submit, on a voluntary basis, each new chemical to the government's scientific subcommittee for approval. The manufacturer provides data on the chemical and physical properties of its candidate pesticide, as well as toxicity data developed through experiments on mammals, birds, fish, and insects. Government advisory scientists make a critical study of the manufacturer's data, as well as an analogy with related chemicals. On this basis they make their recommendations of approval or disapproval. The industry always has abided by these recommendations.

"It's really an organized muddle," one government official has said. "It's a *muddle* because it looks like one and *organized* because it works rather well."

Nevertheless, the chemical industry in Great Britain, through the Association of British Manufacturers of Agricultural Chemicals, has lately moved to have the scheme made mandatory. The manufacturers fear that a new firm might blacken the industry's name by peddling whatever poisons it pleased.

"We would like the voluntary scheme to be made mandatory," one manufacturer has said, "because it would relieve us of responsibility."

The British government seems to hold most of the right opinions although, like its counterparts almost everywhere, it sometimes fails to grasp the enormity of the pesticide problem. The Ministry of Agriculture distributes a booklet, *Chemicals for the Gardener*, which attempts to be helpful about pesticides for the enthusiastic amateur. The producers of the booklet have missed the point of such help. They seem to feel that it lies in recommending a deluge of pesticides for every minor annoyance that the home gardener is likely to encounter. Every insect, every unwanted weed or grass, has its deadly antidote here, and sometimes two or three.

"I find it a rather frightening compilation," writes Dr. Kenneth Mellanby, the entomologist who is Director of the Monks Wood Experimental Station in England.*

*

But in the United States, in the summer of 1963, matters were moving much more slowly than in Great Britain. Rachel Carson was preparing for what was to be her final trip to the coast of Maine. She knew that she was dying. "I seem to have so many matters I need to arrange and tidy up, and it is easy to feel that in such matters there is plenty of time," she told a friend. "I still believe in the old Churchillian determination to fight each battle as it comes, and I think a determination to win may well postpone the final battle."

On the eve of her departure for Maine, she appeared before Senator Ribicoff's subcommittee to offer several recommendations about pesticide use. The following testimony indicates that she was fighting other battles as well as her own.

* Another phase of the problem in Great Britain is discussed in Chapter 9.

1. I hope this committee will give serious consideration to a much neglected problem — that of the right of the citizen to be secure in his own home against the intrusion of poisons applied by other persons. As a minimum of protection, I suggest a legal requirement of adequate advance notice of all community, state, or Federal spraying programs, so that all interests involved may receive hearing and consideration before any spraying is done. I suggest further that machinery be established so that the private citizen inconvenienced or damaged by the intrusion of his neighbor's sprays may seek appropriate redress.

2. In another area, I hope this committee will give its support to new programs of medical research and education in the field of pesticides. I have long felt that the medical profession, with of course notable exceptions, was inadequately informed on this very important environmental health hazard . . .

3. I should also like to see legislation, possibly at the state level, restricting the sale and use of pesticides at least to those capable of understanding the hazards and of following directions . . .

4. I should like to see the registration of chemicals made a function of all agencies concerned rather than of the Department of Agriculture alone . . .

5. It seems to me that our troubles are unnecessarily compounded by the fantastic number of chemical compounds in use as pesticides . . . I should like to see the day when new pesticides will be approved for use only when no existing chemical or other method will do the job.

6. In conclusion, I hope you will give full support to research on new methods of pest control in which chemicals will be minimized or entirely eliminated . . . Since our problems of pest control are numerous and varied, we must search, not for one superweapon that will solve all our problems, but for a great diversity of armaments, each precisely adjusted to its task. To accomplish this end requires ingenuity, persistence and dedication, but the rewards to be gained are great.

The dissenting view at the subcommittee hearings was presented most forcefully by Dr. Mitchell R. Zavon, Associate

Professor of Industrial Medicine at the University of Cincinnati and a consultant to the Shell Chemical Company. "Miss Carson is talking about health effects that will take years to answer," he told the subcommittee. "In the meantime, we'd have to cut off food for people around the world. These peddlers of fear are going to feast on the famine of the world — literally."

During a recess in the hearings, one of the agricultural experts present gave a reporter another slant on what the pest control people thought of the controversy. "You're never going to satisfy organic farmers or emotional women in garden clubs," he said.

Rachel Carson's appearance at the subcommittee hearings marked her last active gesture in American public life. However, she continued to work on through the remainder of the year. Her correspondence was enormous; concerned people wrote from all over America and Europe to tell her of instances of pesticide abuse, or to ask her advice about specific chemicals. She tried to answer every letter and, somehow, found time to write a foreword for Ruth Harrison's book, *Animal Machines*, which told the story of the many cruelties practiced in Great Britain's "factory farming." Rachel Carson's name counted impressively now on books in most of the Western nations; it was linked with a cause that stirred growing numbers of people. Consider this scene that took place during the fall of 1963 in The Netherlands.

Dr. Briejèr, who had championed Rachel Carson's cause in his own country, sat in the living room of his home in Wageningen watching television. The date was November 22. He had been waiting for the start of a program called "Focus," which came on after the news, and which that evening would carry an interview with Dr. Briejèr about pesticides. Suddenly the regular news summary was interrupted by a bulletin from Dallas: President Kennedy had been assassinated.

After a series of further bulletins, the evening's regular programming continued. "Focus" appeared with its scheduled sequence of topics — a short report on French politics, and then the scene changed and Briejèr watched films of crawling insects, spray equipment in operation, and finally a shot of himself, sitting behind a desk. The interviewer was reading from the Dutch translation of *Silent Spring*. There was a closeup of the book, with an inscription in it, dated August 19, 1962, from Rachel Carson to Briejèr. The program went on to a discussion of the PSAC Report, and then to a report on a new Dutch pesticide law which had taken eight years to pass. And then it was over, and the program concluded with a report on Yugoslavia before returning for the rest of the evening to the news from Dallas.

Briejèr sat in front of the television set for a long time. The moment he had looked forward to had been blighted. Then, as he told Rachel Carson in a letter soon afterward, "I realized somehow that that program had been appropriate." President Kennedy had appeared in all three segments of "Focus." The scenes had emphasized his impact on all parts of the world.

"Petty officials in this country, which can hardly be detected on the map, are too busy and do not think of the pesticide problem as important," he wrote. "The President of the United States, however, with a burden heavier than anyone can imagine, was not too busy, and he thought it important."

*

Rachel Carson died at home on April 14, 1964. Scientists, conservationists, and government officials attended her funeral in Washington's National Cathedral several days later. Among her pallbearers were Stewart L. Udall, the Secretary of the Interior, and Senator Ribicoff; the largest wreath in the cathedral had been sent by Prince Philip. And on the floor of the United States Senate, Senator Ribicoff paid tribute to "this gentle lady who aroused people everywhere to be concerned with one of the most significant problems of mid-twentieth century life — man's contamination of his environment."

It was only the end of a single chapter in this long struggle.

The critics of *Silent Spring* continued to fight the book from a standpoint of ignorance: "We have no evidence that the persistent pesticides are harmful — so don't stop us from using them."

Yet the debate would never again be the same. Rachel Carson had pointed out that evidence.

Part II

The Circle of Evidence

7. Miss Carson's "Nightmares" Unfold

1,2,3,4,10,10-hexachloro-6,7-epoxy-1,4,4a,5,6,7,8,8a-octahydro-1, 4-*endo-endo*-5,8-dimethanonaphthalene.

This is the chemical name of endrin, an "economic poison" used in agriculture against soil and foliage insects. Its chemical name may provide a clue to its complexity (complexity of name and composition are qualities endrin shares with most other synthetic, organic chemical pesticides) but does not suggest its toxicity. Here is what Rachel Carson said in *Silent Spring*: "Endrin is the most toxic of all the chlorinated hydrocarbons. Although chemically rather closely related to dieldrin, a little twist in its molecular structure makes it five times as poisonous. It makes the progenitor of all this group of insecticides, DDT, seen by comparison almost harmless. It is 15 times as poisonous as DDT to mammals, 30 times as poisonous to fish, and about 300 times as poisonous to some birds."

Chemists have learned to create these compounds with marvelous facility in their laboratories. They also have "tested" the hazards of their compounds on a great many laboratory specimens. Then, confident of both the effectiveness and the safety of the finished products, they hand them over to the salesmen who, in turn, dutifully peddle them to the world. The process would be foolproof, but for the forgotten element: the chemists, who have solved their special laboratory problems, set these compounds loose in the immensely complex living world, whose problems the biologists have not yet solved.

All the great technological hazards of our time have been

exposed, not in the laboratories, but in the world around us. Stable detergents, nuclear fallout, smog (created partly by the "beneficial" automobile) and thalidomide were each thought at one time not to pose special problems to human beings or their environment. Barry Commoner, speaking of these contaminants, has said: "In each case the risk was undertaken before it was completely understood. The importance of these issues to science, and to the citizen, lies not only in the associated hazards, but in the warning of an incipient abdication of one of the major duties of science — prediction and control of human interventions into nature."

Once loose in the environment, certain chemicals are difficult to control because of their baffling complexity. Even when the harm has been noted, the sources of contamination are difficult to detect. Consider the case of the "Mississippi Fish Kill," and the work of a young scientist named Donald I. Mount who has brought us a big step along the road toward an understanding of the damage that even infinitesimal traces of toxic chemicals can cause in the environment.

*

On November 18, 1963, Robert LaFleur of Louisiana's Division of Water Pollution Control made a phone call to the Public Health Service in Washington. The call was unusual for several reasons. LaFleur's purpose was to report a fish kill of extraordinary proportions on the lower Mississippi River. His call was made despite a general reluctance by state agencies to seek federal help in dealing with matters of environmental contamination; state officials resent federal "interference," feeling that it reflects on their own competence.

But the state of Louisiana had no choice in that situation. It was confronted with a widespread and potentially dangerous problem with which it could no longer contend alone. An estimated five million dead fish were floating, belly up, in the great river which drains one third of the United States, provides drinking water for over one million people, and supports a substantial portion of this country's fishing industry.

One such kill, even one of this magnitude, might have been attributed to natural causes and shrugged off. But fish kills were becoming an annual event in Louisiana along the Mississippi and Atchafalaya Rivers and over hundreds of miles of related bayous. The kills (which sometimes included snakes, eels, and turtles) had begun in the late 1950's. At least thirty large fish kills were reported to the authorities during the summer of 1960. "The sports public, indignant citizens, harassed elected officials and the press were all demanding that something be done," Kenneth Biglane, who was then the chief of the Louisiana Water Pollution Division, has said. "Dead fish were observed to be clogging the intake of the Franklin power plant and fish were dying in Bayou Teche, a stream used as a source of drinking water for the town of Franklin."

In November the kills reached "epidemic" proportions. Although numbers of carp, threadfin shad, and freshwater drum were found dead, catfish composed 95 per cent of the kill. This was particularly disturbing to the authorities because catfish are the principal source of food for many of the poorer people who live in the river towns and bayou settlements. State officials combed the river for clues. Temperature changes in the water, its alkalinity, and its dissolved oxygen content were analyzed and found normal. Abnormality could be detected only in the fish themselves.

"Most of the catfish were bleeding about the mouth," Robert LaFleur later testified, "and many were bleeding about the fins. In every instance examination revealed that this was due to distention of the swim bladder and the digestive tract. The latter was devoid of food material and contained only gas and a small amount of bile-like frothy material. Analysis of the bottom organisms revealed that an abundant food supply was available. Dying fish were swimming at the surface, often inverted, in a very lethargic manner, and were easily captured by hand."

Experts in fish diseases from the U.S. Fish and Wildlife Service were among those who examined the dying fish. No significant number of pathogens was found. Nevertheless, when the kills finally subsided, the state tentatively attributed them to "abdominal dropsy." Smaller fish kills which hit the rivers and bayous in 1961 and 1962 accordingly were written off as "natural." Then, in 1963, another massive kill occurred. Brackish water fish such as menhaden, mullet, sea trout and marine catfish joined the freshwater species as victims. It was at this point that the state had second thoughts about "abdominal dropsy," and turned to the U.S. Public Health Service.

Scientists from the Division of Water Supply and Pollution Control* were dispatched to Louisiana. Before they even approached the water itself, they came across indications of the extent of the fish kill. Sportsmen told of ducks, chiefly fish-eating species such as mergansers, floating lifeless in the streams and bayous. Commercial fishermen had given up in despair, taking jobs in the off-shore oil industry while their wives went to work in canneries. Wholesalers complained there were hardly any fish to buy.

"You can stand here on the dock any day of the fall and winter and see thousands of dead fish float by with the current," one of the wholesalers told a reporter from the *New York Times*. "They would shoot up out of the water and just flop over. Many others died in the nets before the fishermen could bring them in."

Among the federal scientists on the scene was Donald I. Mount, a young biologist attached to the Public Health Service.

* This division, a part of the Department of Health, Education and Welfare, was abolished by the Water Quality Act of 1965. The Act created the present Federal Water Pollution Control Administration (FWPCA), effective January 1, 1966. Four months later, under the President's Reorganization Plan, the FWPCA was transferred to the Department of the Interior.

After discussing the pattern of the kills with state authorities at Baton Rouge, Mount and another biologist set off to investigate conditions on the river.

"We saw dying fish in the river and canals in the Baton Rouge area and in the passes of the Mississippi River near the coastal town of Venice," Mount has written. "Channel catfish, drum, buffalo and shad were most affected, but we also observed acres of minnows at the surface that would convulse when stressed by our boat. In the brackish water area, mullet and menhaden were observed jumping several feet out of the water when disturbed by the wake of a boat; frequently we saw them land on the levees or oil well platforms."

Mount was struck by the similarities exhibited by all of the dying fish: convulsions, loss of equilibrium, some hemorrhagic areas on their bodies, and surface swimming. When he collected and examined them, he found that almost without exception their entire tracts were devoid of food, even in trace amounts. Yet there were excessive deposits of visceral fat in the fish, and their length/weight ratios were considered normal.

Mount and his colleague packed 100 pounds of dead fish in dry ice, preserved the internal organs of other fish, took blood samples and smears, and collected and froze stream bottom sediments from areas where the kills had occurred. Part of the collection was shipped to the U.S. Fish and Wildlife Service Fish Laboratory at Leestown, West Virginia; the bulk of it was sent to the Public Health Service's Taft Sanitary Engineering Center in Cincinnati, which was Mount's home laboratory.

Then began weeks of laborious analysis to track down, first, the killing substance and, later, its source. A report from the Fish and Wildlife Service arrived in Cincinnati: there was no evidence of disease or parasites in the dead fish. Progressing through a series of examinations, Mount and his colleagues discarded, one after the other, the possibilities of botulism, viruses, and toxic concentrations of metals. Their studies of blood smears disclosed low counts of both red and white cells, which suggested that disease was not the cause of death. The suspicions of the workers were aroused when tissues of the fish were fed to laboratory mice, bringing on symptoms of toxicity among them. But the tests were inconclusive.

A major breakthrough occurred when the Public Health Service scientists turned to the mud they had dredged from affected areas in the river. Chloroform extract residues of these bottom sediments, when dissolved in tanks where healthy catfish swam, caused dramatic reactions. The healthy fish convulsed, lost their equilibrium and hemorrhaged — symptoms which corresponded exactly with those observed among stricken fish in the Mississippi and Atchafalaya Rivers!

Mount and his colleagues followed up this lead. Steam distillates, taken from the livers of the dead fish, also were found to bring on the familiar symptoms among healthy fish. The evidence now suggested that some foreign toxic substance, not a part of the water's normal flow, had affected the fish. Yet, despite the dramatic result of their investigations, Mount and his colleagues realized that their work had only begun. The analysis had been more thorough, perhaps, than those undertaken after previous fish kills, but very little scientific evidence had been amassed to back up the long-held suspicions of many biologists along the Mississippi that they had encountered a crisis brought on by the chemical contamination of the environment.

But Mount and his colleagues held several advantages over their predecessors. Only very recently had instruments so sensitive been developed that they are able to measure chemical residues that occur in traces as slight as one part per billion (and now in parts per trillion!). Further, Mount devised new techniques of "fish autopsy" which enabled him to push toward a resolution of the problem.

Teams of scientists at Taft and other PHS laboratories using machines called gas chromatographs analyzed extracts of both the fish tissues and the bottom sediments. Traces of both DDT and dieldrin occasionally were found, but not in significant amounts. Yet three "peaks" were common to the electron-capture gas chromatograms of all the extracts. One of the peaks was thought to be endrin, a chlorinated hydrocarbon insecticide that was used extensively in Louisiana to control the sugarcane borer. But the other two substances whose existence was revealed by the peaks on the chromatograms could not be identified. The scientists referred to these unknown substances as X and Y.

"All attempts to identify X and Y, including infrared analyses, failed," Mount wrote. "Finally, however, we devised a procedure whereby the material producing each gas chromatographic peak could be collected in nearly pure form and assayed using new-born guppies in 1 to 2 milliliters of water. These assays revealed that, from a given quantity of tissue, only the material for the peak thought to be endrin was toxic, and it produced in the guppies the well-known symptoms of endrin poisoning."

Chemists carried on one phase of the investigation at this point, trying to pin down the identity of the substances which had appeared on the gas chromatograms. Through infrared analysis, the toxic peak was identified unequivocally as endrin. Later, with the help of chemists from the Shell Chemical Company (one of the producers of endrin), the X and Y peaks were identified as chemicals associated with endrin's manufacture.

At this point the nonscientist might conclude that the culprit stood convicted, but that was by no means the case. Circumstantial evidence, no matter how suggestive, cannot be accepted as scientific proof. Cause must be tied firmly to effect. A diagnosis of death through pesticide poisoning, even though relatively high levels of pesticide residues have been discovered in the victim's tissues, is not necessarily valid. The variety of chemical components, and their actions on living creatures, are immensely complicated. Consider this report made by two USDA researchers in 1960:

> In feeding tests, using a diet containing 60 parts per million of heptachlor, 52 parts per million of heptachlor epoxide could be recovered from the fat of cattle at the end of a 16-week feeding period. The cattle at that time were in excellent health and condition . . . By contrast, a calf poisoned and killed by a single large dose of heptachlor, revealed a residue of only 2 parts per million in its fat.
>
> In our feeding trials with lindane, using a level of 100 parts per million in the feed for 10 weeks, a residue of 100 parts per million existed in the fat at the end of the feeding period . . . By contrast, three cattle of similar breed, age and condition, deliberately poisoned by high concentrations of lindane in dips, revealed only 23 parts per million 1 week later in their fat.

Scientists have learned that, even among species of fish or animals that are closely related, there is a wide divergence in their susceptibility to chemical pesticides. Dosages, or residue levels, which might prove fatal to one species, apparently have little effect on others. And, of course, such variants as the nature of the specific chemical, the normal diet of the victim, and the victim's physical condition at the time it consumed the chemical, all figure significantly in the ultimate toxic effect.

"Fortunately, the problem could be subjected to experimental study," writes Lucille F. Stickel of the U.S. Bureau of Sport Fisheries and Wildlife, "and it was learned that some tissues lend themselves to diagnostic interpretation, whereas others do not. Subsequently, diagnostic tissues and residue levels indicative of DDT poisoning have been established that apply across a wide range of animal species. Experiments have also been conducted to determine critical levels of certain of the DDT metabolites. Diagnostic tissues and residue levels have been established for dieldrin, and these also apply across a wide range of animal species."

It was Donald Mount's triumph that he established the levels of endrin in the blood of channel catfish, buffalo and gizzard shad that invariably prove fatal. The blood samples taken from stricken fish in the Mississippi River all contained endrin residues substantially above the "fatal" levels. In no case, after the kills had subsided, did fish collected in the river contain pesticide residues approaching these levels.

What was the source of the endrin in the river? Although it had been known for a long time that persistent pesticides like endrin often were washed from fields into nearby rivers, neither state officials nor Public Health Service investigators on the scene could trace large amounts of contaminants in the river to the sugarcane fields. The biggest kills were occurring late in the fall, while the fields were sprayed in the spring. Furthermore, most of Louisiana's cane fields are separated from the Mississippi and Atchafalaya Rivers by levees. One state sanitary engineer summed up the investigation this way: "Although the concentrations seem to increase in reaches of the Mississippi River in Louisiana, there is really no drainage to the river from Louisiana soils and there are no Louisiana industries discharging insecticide wastes into the river."

Louisiana officials concluded that most of the toxic substance affecting local fisheries must have been carried into the state in the river's flow from the north. It was on this basis that they had requested the assistance of the Federal Government, and it was on this basis that the investigators proceeded once endrin had been identified as the killer. Their task was not easy. There are about 100,000 industries along the Mississippi, many of which pour their wastes into the river and its tributaries. Over a hundred of these factories manufacture pesticides. The complexity of the Mississippi's wastes boggles the imagination.

"Stream scientists working for Louisiana's Division of Water Pollution Control," Kenneth Biglane, its former chief, has said, "soon become educated to the different types of water pollution that are found in their state from time to time. Wastes from sugar factories, sweet potato canneries, pulp and paper mills, oil field brines, naval stores plants, chemical plants, municipal sewages, and slaughterhouses all have two things in common. They can degrade water and they can kill aquatic life. Their point in time, their point source, and their physical and chemical alteration of the aquatic environment, however, offer clues to their dissimilarities."

The Public Health Service investigators had come upon several clues. From farmers and biologists, they learned that fish, trapped in endrin-infested streams near sugarcane fields immediately after spraying and heavy rains, usually died at once. This effect they described as "acute." But the more widespread kills, such as that which had occurred in the fall of 1963, were accompanied by symptoms which the scientists described as "subacute." The fish died slowly, after passing through a number of symptoms; or else they suffered convulsions, righted themselves, and survived. The investigators reasoned that the fish may have picked up endrin in smaller quantities, after it had been diluted by the river's immense flow. At Memphis, Tennessee, some 500 miles up the river, there was a plant which manufactured endrin.

In April, 1964, Alfred R. Grzenda led a team of pesticide experts from the Public Health Service to the Memphis plant, which was owned by the Velsicol Chemical Corporation of Chicago. This was the corporation which had tried to stop the publication of *Silent Spring* in the summer of 1962. Endrin, in fact, had been developed in Velsicol's Chicago laboratories.

Grzenda's job was to gather information on the manufacture of endrin, and collect samples of mud and water in the vicinity of the plant. Apparently the government investigators did not receive a very warm welcome. When Grzenda asked specific questions about endrin and its by-products, the plant's manager referred him to the patents for their manufacture. In their search for samples, the investigators were treated more like the agents of an alien government than members of the United States Public Health Service.

"The Velsicol Company more or less selected the sites which

we sampled," Grzenda said. "In other words, it was a guided tour through the plant."

Grzenda and his team investigated various sewers in the area and discovered, in *all* of them, strong chemical odors which he said were similar to those noticed around Velsicol's newly installed waste treatment plant. They also found a bypass sewer line which discharged Velsicol's trade wastes into a lake which flowed into Wolf River, a small tributary of the Mississippi. And, at some distance from the factory, they found a place called the Hollywood Dump, where Velsicol's solid and semiliquid wastes were hauled daily in large containers.

"This material is buried, or left exposed on a portion of the dump located on the flood plain of the Wolf River," Grzenda said. "All of the sites are subject to flooding. The plant manager denied that any solids from the endrin plant were being hauled to the Hollywood Dump, but we noted drums labeled 'isodrin scraps' at the dump on April 15. Isodrin is one of the compounds used in the manufacture of endrin. However, the manager said that such material is not normally taken to the Hollywood Dump for disposal."

Samples of water taken from the sewers and waterways around the Velsicol plant revealed quantities of endrin, exceeding all previous reports of concentrations of chlorinated hydrocarbon insecticides in water. According to the PHS report, most of the contaminated water was "known to be downstream from points used or previously used by the Velsicol Chemical Corporation for waste discharge or disposal."

And the complete PHS report, made public in the spring of 1964, spoke in more specific terms than one ordinarily finds in a government report: "Endrin discharged in the Memphis, Tennessee, area, other sources of endrin not yet identified, and possibly other pesticides and discharges of sewage and industrial wastes of many kinds, pollute the waters of the lower Mississippi and Atchafalaya rivers and thereby endanger health and welfare of persons in a State or States other than those in which such discharges originate. Such discharges are subject to abatement under the provisions of the Federal Water Pollution Control Act."

The United States Public Health Service felt no strong compulsion to charge into the pesticide-public health controversy. PHS always has been reluctant to undertake administrative functions, particularly those which include enforcement activity. For example, it did not prod state health departments to enforce local pollution laws, believing that such pressure by a federal agency "would destroy state relations and dry up future research." It was chiefly for this reason that the Federal Water Pollution Control Administration finally was removed from PHS influence by the President's reorganization plan and transferred to the Interior Department in 1966.

But in 1964 PHS was drawn, like so many other agencies, organizations, and politicians, into the great whirlpool created by the revelations of its investigators. Since water purification systems do not remove these complex chemicals (unless special processes, such as activated carbon treatment, are applied), traces of endrin had been detected in the drinking water of New Orleans and Vicksburg. Dr. James M. Hundley, the U.S. Assistant Surgeon General, had expressed misgivings about the health of those people in low income groups who subsisted largely on a diet of catfish.

Official Washington was in an uproar. Secretary of the Interior Stewart L. Udall reacted forcefully to the increasingly obvious menace of persistent pesticides. In an order to his chief assistants which established a new departmental policy, Udall said: "It is essential that all pesticides, herbicides, and related chemicals be applied in a manner fully consistent with the protection of the entire environment. Any contemplated use of these chemicals must take into account both known and possible environmental effects. The guiding rule for the Department shall be that when there is a reasonable doubt regarding the environmental effects of the use of a given pesticide, no use should be made."

The U.S. Department of Agriculture did not respond quite so forcefully. Though its survey of pesticide plants along the river found conditions "which appear to constitute a definite hazard," it said that the large quantities of endrin found in the river around Memphis did not result from agricultural use, and did not require action on its part. It called, not for any legislative or departmental action, but for "further study."

Senator Abraham Ribicoff scheduled a new round of his subcommittee hearings because of the Mississippi fish kill. (Rachel Carson, who heard the news shortly before her death, told a friend she was "delighted" by Ribicoff's action.) The hearings served as another forum for the struggle between opposing forces.

Perhaps at the heart of the dispute lay the inability of many pesticide-oriented businessmen and officials to believe that comparatively tiny quantities of any poison in a great river could bring about a major disaster. Donald Mount has spoken of those "who, seeking the truth, fail to comprehend the magnitude of endrin toxicity to fish and other aquatic animals. Perhaps it *is* difficult to understand that any substance in water in such minute concentrations as 0.1 parts per billion could be acutely toxic to fish. However, one must consider that in just two hours, the blood of a catfish can attain an endrin concentration of 1,000 or more times greater than that of the water in which the fish swims; understanding then comes more readily.

"In one of our studies we discovered that fathead minnows exposed to .015 parts per billion, had total body concentrations 10,000 times greater than that of the water. Because of such concentrating ability, it is obvious that accurate toxicity data cannot be obtained when one or more pounds of bullheads are placed in a five- or ten-gallon aquarium in which the test water is *not* renewed continuously . . . one can readily realize the need to measure concentrations of endrin in the parts per billion range."*

* Rachel Carson, in *Silent Spring* (page 41), had noted a tendency of persistent

The report of Ribicoff's subcommittee, when it was published, took a critical look at both industry and the various government agencies involved in the dispute. Of the industry representatives, the report said, "Their testimony in the hearings dealt mostly with their accomplishments and there seemed to be little recognition of ecological problems."

And of government agencies, the report had this to say:

"The Government is not properly organized to get at the facts when a complicated technological problem affecting public health attracts wide attention in the press. Departments of the Government had inherent conflict in the fish kill situation as shown by their mission responsibilities.

"The jurisdictional dispute extends even to agencies within departments. Information is withheld or leaked to support bureaucratic positions rather than contribute to the body of scientific understanding. Scientists retreat to positions of technical 'nit picking' (such as the limitations of infrared absorption spectra for qualitative organic analysis) which are of little help to a congressional committee. Industry, urged on by fears of its image to the customer and further Government regulation, abdicates a responsibility for industrial waste which is inescapable. Local governments display technical ineptness and unbecoming chauvinism."

The Secretary of Health, Education and Welfare, Anthony J. Celebrezze, later convened a conference in New Orleans on Louisiana's call for the abatement of interstate water pollution. The conference itself was occasionally marred by the pouting and grumbling of various state officials, who resented federal "intrusion" in a crisis they could not deal with themselves. And,

---

pesticides to "disappear" from a lake's water, but to reappear later in "the fabric of the life the lake supports." Elsewhere, George J. Wallace and Ernest A. Boykins have described a similar phenomenon in soil: "Since soil residues decline, and accretions in earthworms and presumably other soil organisms do not, we suggest that much of the so-called 'disappearance' of persistent chemicals is really transfer and re-distribution — from soil to soil organisms, and then to higher animals." Some of the optimistic reports based on the monitoring of water and soil are, obviously, of a very dubious quality.

on the floor of the United States Senate, Senator Everett McKinley Dirksen of Illinois rose to condemn the Public Health Service, claiming that it had made "wild accusations" and had "unjustly crucified" Velsicol before collecting all the facts.

Later, Dirksen inserted into the *Congressional Record* quotations from several of the academicians who had criticized *Silent Spring*; the gist of these remarks was that it was not the pesticides which posed the danger, but those who criticized their use and therefore exposed the world to starvation and disease. Dirksen's support of Illinois-based Velsicol surprised no one; he had always been in the forefront of those who had tried to defend industry against effective legislation to control the manufacture of potentially dangerous drugs.

A beginning had been made in the effort to unravel that fantastic tangle of chemical threads we have come to call "the pesticide problem." There are more than 200 basic chemicals used in modern pest control. They are combined over and over so that the pest control people have at their fingertips more than 60,000 separate compounds. Each of them reacts differently, under different conditions, on different individual organisms. It should not be wondered at that there remains some doubt about whether all of the chemical sources of the Mississippi fish kill were detected. The wonder is that the government investigators were able to pick out of the river's dense web of pollutants the single most hazardous substance and trace it to its source.

In this respect, the Mississippi disaster ended in a queer sort of triumph for the government investigators. They kept Velsicol's Memphis plant under constant surveillance. Its discharges were sampled daily and statistics on the level of pesticides in nearby waters were carefully recorded. Occasional peaks registered on the monitoring devices suggested that contaminants still found their way to the river.

"But the round-the-clock surveillance by our expert technical staff eventually paid off," says Murray Stein, who was active in the Velsicol matter and who later became Assistant Commissioner for Enforcement of the Federal Water Pollution Control Administration. The peaks eventually disappeared from the government's monitoring devices. Unabashed by the past, Velsicol's public relations arm renewed its attack on *Silent Spring*: "In case you haven't noticed, trees leafed, birds sang, squirrels reconnoitred, fish leaped — 1965 was a normal spring, not the 'silent type' of the late Miss Carson's nightmares."

Soon afterward the Velsicol Chemical Corporation came under new ownership, and Stein and his colleagues noticed an abrupt change in the company's policies. "They came to see us in Washington, and later we met with them in Chicago," Stein says. "They listened carefully to our story. Since then, they have been very cooperative with us."

Because the Federal Water Pollution Control Administration does not have subpoena power, its investigators had never been able to undertake thorough inspections of the Velsicol plant. This situation altered under the new ownership, and the government people have been invited inside to look around. Velsicol also sealed up the residue-clogged sewers, and bought a large farm nearby, where its employees can bury empty pesticide drums.

Is the government satisfied with its case?

"Until we went after the Velsicol people, the fish kills were persistent," Stein says. "Once the discharges stopped, the big fish kills stopped. Draw your own conclusions."

8. Tainted Waters

THE MISSISSIPPI RIVER FISH KILL fulfilled the government's need for a clear-cut case: plotted, as if on a graph, neatly labeled, and filed away with a sense of satisfaction. It was, basically, a classic detective story. Yet behind the untangling of a single poisonous thread in the great river lay a more unpleasant — and more important — truth. Neither law nor science can restrict any longer the global currents of contamination. In thousands of lakes and rivers the unseen persistent poisons accumulate, tainting whatever life they touch on their journey toward the sea. The fatty substance of living things is their true element. Found in the water only in the most minute traces, the poisons seek out and insinuate themselves into the surrounding life, to build up through the food chain to frightening proportions.

• In Michigan, the state's promising coho salmon program was dealt a severe blow in 1968 when over 700,000 salmon fry died in hatcheries. Their death was attributed by biologists to DDT, which had drained or drifted into the water from farmlands.

• In Canada, where watersheds had been sprayed with DDT in June, wild young salmon were found dead in streams the following autumn. The DDT residues had been stored for months in their tissues without conspicuous effects. When falling water temperatures forced the fish to call on their fat reserves, the residues suddenly were redistributed throughout their bodies and into vital organs.

• In Maine, a fisheries expert predicted that persistent pesticides, carried to the sea by rivers from inland farms and forests, soon would damage the state's famous lobster fisheries. "If the practices of the past are continued," he said, "there will be a strip of inshore waters that will be intolerable to the lobster." Traces of DDT were found even in lobsters taken 100 miles at sea.

• Various marine fish, including hake, mackerel, and tuna, now contain DDT residues in excess of those found in freshwater fish in many pesticide-contaminated lakes and rivers. Biologists fear that these marine fish may experience the same trouble producing viable fry as do their freshwater relatives.

These unsettling disclosures have come at a time when the world's nutritionists are looking toward the sea in their attempts to close the "protein gap." Obviously the world's exploding population soon will strain the land's capacity to feed it. Not even the "green revolution" — in which modern machinery and new plant strains are increasing food production in the underdeveloped countries — will be able to solve all of our food problems. New sources of food must be found — and found quickly. Man has turned with renewed interest to an investigation of the sea.

Yet even the wide blue sea is not an inexhaustible larder. (Fishermen know that the bluer the water's surface, the more barren its depths.) Most of the sea's harvest is taken from the continental shelves, while the great ocean basins themselves are as infertile as a desert. The 130 million pounds of food we take from the sea each year is already a considerable portion of the available harvest. Here and there we find evidence that this food resource is overfished and depleted.

Confronted by the knowledge that this precious resource must be protected, man is yet in the process of jeopardizing it by the use of those pesticides he hopes will *increase* his food supply. There was a time when he might have pleaded ignorance. As late as 1963, the United States Bureau of Commercial Fisheries was most reluctant to release any information about residues found in deep sea fish. Instead, the bureau claimed there were no data to justify the reports of such residues. But in that same year, in his testimony before the Ribicoff Committee, Secretary of the Interior Udall cut the ground out from under his own bureau by citing such information. (Needless to say, the testimony was prepared elsewhere in the Department.) Since that time, the bureau has faced up to the unpleasant facts.

It is apparent today that the persistence of DDT and the other chlorinated hydrocarbon pesticides is especially insidious in the sea. These poisons are greatly magnified as they pass up through food chains. Persistence itself is not necessarily objectionable, and may even be an asset to a poison under certain circumstances. The objection to the chlorinated hydrocarbons, as Rachel Carson pointed out, is not so much that they remain remarkably stable over a period of years but that they "magnify" in living tissues. As they move up a food chain, from plants through smaller herbivores to a succession of larger and larger carnivores, the poisons become more concentrated in living tissues.

Most of the domestic animals and wildlife that man uses for food are herbivores. Thus they are low on the food chain, not having had the opportunity to consume the fat (and the pesticide residues it contains) of other animals. Most seafood, on the other hand, is carnivorous. Even the herring is several stages away on the food chain from the plant life at the base of the nutrient pyramid.

Yet the immediate fear is not that man will consume sufficient residues in his food to poison himself (although that possibility must be considered at some point). The immediate danger in this hunger-haunted world is that the food chain itself, from its base up through the larger carnivores, will become impaired. Let us look at some of the possibilities.

American scientists have traced a step-by-step increase in DDT through a Lake Michigan ecosystem. The lake is contaminated chiefly by aerial drift from spraying operations. Concentrations of DDT in bottom muds averaged 0.014 parts per million; averaged 0.44 ppm in shrimp, more than ten times as high; leaped another ten times to 4.5 ppm in chub and 5.6 ppm in whitefish; and finally to 98.8 ppm in scavenging herring gulls. The herring gulls, top carnivores (the final stage of the food chain) in this ecosystem, contained levels of DDT about 7,000 times greater than those found in the mud.

These Lake Michigan gulls showed poor breeding success. Between 30 and 35 per cent of the eggs laid in nests surveyed by scientists did not hatch. The eggs themselves contained high residue levels of DDT (or DDE, a toxic breakdown product of DDT). Yet the complexity of assessing pesticide damage is demonstrated again here because the local gull population continues to *rise*. The accumulating pollution of the lake by human wastes and garbage is favorable to the gulls' habitat, thereby offsetting, at least temporarily, the decline in reproduction.

Bernard Venables, a British angler and writer, has remarked that even those rivers which are not yet dramatically polluted have a *thinner* look about them. The banks no longer glow with a variety of flowering plants, the fish themselves are no longer as glistening or as firm to the touch as they once were. The sickness of the rivers has many causes, not least among them the agricultural wastes. "The insecticides, diluted by the floods," Venables says of British rivers, "do not produce an identifiable killing pollution, but by their partial poisoning thin and reduce the animals on which fish feed and debilitate the fish. The weed-killers imperceptibly reduce those water plants in which otherwise fish-feeding animals would breed and live."

Great rivers drain considerable areas of the continents, carrying not only fresh water but nutrients to the sea. Off the coast, where the waters mingle, occur some of the sea's most fertile regions. But the rivers carry pesticides in their flow too. Since DDT is nearly insoluble in water, it insinuates itself into "biological materials," living and dead. The accumulation of DDT in the life of our estuaries is causing increasing concern among biologists.

Part of the reason for this concern is that DDT is attacking the very base of the ocean's food chains. Sunbeams, falling in the ocean, are trapped by microscopic plants called phytoplankton. Through photosynthesis, the phytoplankton convert the energy held in sunbeams into food. Zooplankton graze on the phytoplankton. Young fish feed on the zooplankton, bigger fish and cephalopods (squid) feed on the little fish, and pelagic birds in their turn feed on them.

Charles F. Wurster, Jr., a biochemist at the State University of New York at Stony Brook and a world-famous authority on the effects of pesticides on wildlife, published the results of his important experiments in 1968. They confirmed earlier reports that DDT inhibits photosynthesis in phytoplankton.

"These single-celled algae," Wurster writes, "are the indispensable base of marine food chains, responsible for more than half of the world's photosynthesis; interference with this process could have profound, worldwide biological implications."

Moreover, since phytoplankton take in DDT and pass it on to higher organisms, it is obvious how this omnipresent contaminant is concentrated through zooplankton and the larger organisms until it comes to rest in the organs and fatty tissues of top carnivores.

Another scientist has enlarged on the implications inherent in the depletion of any part of the world's phytoplankton. Professor Georg Borgstrom, in his book *The Hungry Planet*, reports that it required 15 per cent of all the world's algae to produce the fish harvested by man during one recent year. Moreover, phytoplankton take in DDT and pass it on to higher organisms in increased concentrations. Further studies by the Bureau of Commercial Fisheries recently disclosed that only a fraction of the DDT residue levels found in many estuaries (which catch runoff from farmland) caused 35 per cent to 100 per cent mortality in shrimp, crabs, and finfish. The Bureau's biologists concluded that there has been a significant reduction in the productivity of both finfish and shellfish in important estuarine "nurseries." The organisms which survive pass on the toxins to higher forms.

For a time it was believed that persistent pesticide residues would "get lost" in the vast ocean reaches. Now we know that, though the residues are dispersed over wide areas, they eventually are picked up by living organisms. These residues reach the most remote areas. Part of the residues are spread by the water itself. But recent analysis of airborne particulates over both the Atlantic and Pacific oceans indicates that pesticides are also widely dispersed by the wind. The amount of pesticide fallout over the tropical Atlantic Ocean from the northeast trades has been compared to that flowing into the sea from a major river system.

As a result of this dispersal, biologists report that the contamination of oceanic ecosystems by DDT now is reaching levels approaching those of inland waters. DDT residues are regularly found in fish oils (tuna, shark, etc.); while a British study has shown that cod liver oil now contains residues of DDT and other persistent pesticides. Marine birds, standing at the top of many oceanic food chains, contain higher residue levels than fish. This comes as a chilling disclosure when we recall that shearwaters, the "mutton birds" of the southern seas, often are eaten by human beings. No individual organisms are exempt from this remorseless contamination. DDT residues have been detected in seals and penguins even in the Antarctic.

There has been no more dramatic evidence of the extent to which pesticides have penetrated to the most remote corners of the world than the apparent fate of the cahow (which is also called the Bermuda petrel). David B. Wingate, Bermuda's conservation officer, and Wurster recently investigated this bird's declining reproduction.

"The Bermuda petrel is a wholly pelagic species," Wingate and Wurster write. "It visits land only to breed, breeds only on Bermuda, and arrives and departs only at night. The single egg is laid underground at the end of a long burrow. When not in the burrow the bird feeds far at sea, mainly on cephalopods; when not breeding it probably ranges over much of the North Atlantic."

The Bermuda petrel is a rare bird to begin with, almost having been wiped out when the Europeans invaded its breeding grounds in the seventeenth century. It became, by order of the Bermuda government, the first species to be protected by proclamation in the New World.

Nevertheless, the Bermuda petrel disappeared about 1630. It was not recorded again until 1906, when it was rediscovered nesting on the outer islands. Its population increased during the next fifty years until about 100 of the petrels were known to exist. Naturalists believed that under legal protection and in isolation the petrels would maintain their numbers and perhaps (like the whooping crane) show a gradual increase.

But Wurster and Wingate proved that the case is otherwise. These birds live and die utterly remote from areas where pesticides are applied. Yet they have been found to lay eggs containing DDT residues comparable to those found in the eggs of terrestrial birds such as the peregrine falcon, whose decline has been linked to heavy pesticide residues. The slow, painful increase in the petrel population has been halted. Fewer and fewer eggs hatch each year. Since 1958 this decline has persisted, unexplained by any obvious environmental factor. In 1969, 24 pairs of petrels nested, but only seven chicks (one of which later died) hatched. Wurster and Wingate say unequivocally that if the present rate of decline in reproduction (3.25 per cent) continues, the Bermuda petrel will cease to reproduce by the year 1978.

The story of the Bermuda petrel is a clear warning that something has gone terribly wrong in man's plans to increase his food supply. Other, more vital, organisms may be in similar jeopardy. Man finds himself in the position of the carpenter who is trying to lengthen one end of a board by adding to it the sawed-off nether end. The whole structure eventually may collapse.

There are those who believe that man one day may no longer have to depend on the contaminated oceans to narrow the protein gap. These visionaries look forward to the development of aquaculture, in which seafood will be propagated and raised in antiseptic super-hatcheries. A word of caution is appropriate even here. It was provided by Canadian biologists who attempted to raise brook, rainbow, and cutthroat trout in a hatchery in Jasper National Park. They found it almost impossible to buy trout eggs from commercial or government suppliers that did not contain residues of DDT and its metabolites. Those collections of eggs containing greater residues experienced mortality ranging from 70 per cent to 90 per cent. The Canadians encountered further difficulty when they tried to locate suitable commercial dry feeds for the surviving trout fry.

"All commercial dry trout feeds analyzed were found to contain chlorinated hydrocarbons," they wrote. "Of several ingredients used, only brewers' yeast was found to be almost free of contamination. From operational observations, it would seem that DDT in manufactured trout food is detrimental to the growth of trout raised under hatchery conditions, when DDT and metabolites in the eggs and fry exceed certain levels."

DDT is there, from feed to egg to fry to adult. The story is repeated endlessly throughout the three fourths of the earth's surface that is covered by water. Many of DDT's chemical relatives act in the same manner. Those optimists who seek to solve the world's food problem by harvesting or cultivating the sea may be obliged to revise their plans.

9. "We Have Reached These Conclusions"

SINCE THE END of World War II, government, industry, and the academic world have been curiously slow to react to the hazards posed to the environment by persistent pesticides. The ordinary channels of communication have been clogged by propaganda, sales talk, and irrelevancy. Effective research and understanding have come very slowly indeed. And only in the years since the publication of *Silent Spring* have American and British scientists joined to lay bare convincingly both the effects of pesticide contamination in the environment and its mechanism in the living organism.

*

The mass slaughter of British birds (and scavenging mammals) was halted abruptly by the ban on the offending chlorinated hydrocarbons for spring seed dressings in 1962. Here the effect had been *acute*. But British scientists, aware by then of the potential *chronic* effects of these pesticides, continued their investigations. Inevitably, more insidious effects were uncovered. The scientific work produced by these investigations has proved to be among the most interesting and alarming in the entire field of pesticide research.

Life has not been at all secure for the large birds of prey in England since that distant time when gentry of ancient lineage and broad acres assigned to their gamekeepers the task of waging war against creatures (human or otherwise) which preyed on their game. Down to recent times the gamekeeper's gibbet has been a prominent feature of the rural English landscape. Edward Thomas described some of those wild victims of the gamekeeper's gun:

> And many other beasts
> And birds, skin, bone and feather,
> Have been taken from their feasts
> And hung up there together,
> To swing and have endless leisure
> In the sun and in the snow,
> Without pain, without pleasure,
> On the dead oak tree bough.

Birds of prey were mercilessly hunted down during the nineteenth century. In Scotland, when intensive sheep farming took the place of cattle and crofting, hawks and eagles fell even lower in repute since they were considered to prey on lambs. Egg collectors took their toll, too. During World War II, the Air Ministry destroyed peregrine falcons for their inability to distinguish the homing pigeon from the shiftless, pestiferous varieties of cities and farmlands.

In the 1950's, however, it appeared at last that the hawks were to be morally rehabilitated. The great "pests" were discovered to be the herbivores rather than the carnivores. Grain-eating rodents (such as mice and rabbits) and leaf-eating insects were recognized as the farmer's real enemy, and the hawk was seen as among man's best friends. The Protection of Birds Act of 1954 and subsequent government orders placed all birds of prey under protection in Great Britain. Such measures, added to the rapid recovery of persecuted species of hawks after the war, indicated that these magnificent creatures had passed out of danger.

Suddenly, naturalists realized that something had gone wrong. The peregrine falcon, the kestrel, and the sparrow hawk were found to be in the throes of a remarkable decline. The peregrine population, for instance, had dropped more than 50 per cent from its "normal" prewar figure of 650 breeding pairs; only a quarter of the remaining pairs continued to breed successfully. Pesticide residues were found in both the bodies and the eggs of these flesh-eating birds. As J. A. Baker wrote of the peregrines, "Foul poison burned within them like a burrowing fuse."

In western Scotland, the golden eagle plunged into a decline described as "catastrophic" by naturalists and attributed in part to dieldrin sheep-dips (the golden eagle often feeds on the carcasses of sheep which have died natural deaths).

Similar declines were noted elsewhere in Europe and around the Mediterranean among the birds of prey. Twenty of the thirty-one species of predatory birds found in France were declining, and twenty of the thirty-three in Italy. The story in Israel was especially dramatic. In this small but intensively farmed country, the *human* population contained DDT residues in its fat amounting to 19 parts per million (compared to 12 parts per million in the average American and only 2 parts per million in the average Englishman). Evidence of devastating losses from secondary poisoning among Israel's rodent-eating hawks was uncovered by scientists. Wheat seeds, coated with thallium sulphate, were planted in Israel's fields.

"Within ten years," writes Professor H. Mendelssohn of the University of Tel Aviv, "some species like the black kite, the long-legged buzzard, the Egyptian vulture and eagles of the clanga-pomarina group were reduced to less than ten percent of their original population density. Several days after a field mouse-poisoning operation, the birds were found paralyzed or dead in the fields. Birds, paralyzed by feeding on thallium-poisoned mice, were not able to recover in the field as — owing to their being unable to hunt — they slowly starved to death. In case of rain, they were drenched and then their suffering was shortened, as they died of exposure before dying of starvation."

In 1969 Sweden became the first nation in the world to take

sweeping action against the persistent pesticides. The Swedes were jarred into action by the public revelation that DDT had appeared there in human milk. The intake of DDT in breast-fed babies was said to be in the range of exposure at which laboratory animals begin to show biochemical changes. The National Poisons and Pesticide Board forbade the use of DDT and lindane in household and garden use, and banned aldrin and dieldrin for all uses. Further, since investigations had turned up the striking estimate that the amounts of DDT found in Swedish soils exceeded the total quantities ever used in that country, the board also banned the use of DDT outside home and garden for two years. The moratorium was to enable scientists to determine how much DDT is carried into Sweden by winds and rain. Many scientists doubt that DDT's use will ever be permitted in Sweden again.

Mercury remains a problem in Sweden. Mercury compounds are used extensively as fungicides; corn seed, for instance, is treated before planting with methylmercury dycyandiamide. Recent research there suggests that mercury not only presents a grave hazard to wildlife in Sweden, but ultimately may prove equally as hazardous to human beings. Definite proof of these effects has been slow to come to light for several reasons. One is that mercury is found in tiny but natural amounts in all human beings. What is the level at which mercury evolves from a vital component of our bodies into a menace? Another obstacle to defining the mercury problem is the fact that it is widely used in Swedish paper mills and chlorine alkali factories as well as in pesticides. The sources of the omnipresent contamination are therefore extremely difficult to detect, but it is hazardous, whatever the source. Mercury has affected both the physical and mental health of workers in certain industries for more than a century. Alice's friend, the Mad Hatter, was probably a victim of mercury poisoning; the mercury applied to material used in the manufacture of felt hats affected so many workers that the phrase "mad as a hatter" came naturally to

Nineteenth Century tongues. The Swedes recently solved part of their problem by banning the use of one extremely pervasive mercury compound, methylmercury, and shipping it to the United States, where it is now used to treat wheat seed.

*

Today, in the face of accumulating environmental disasters, British scientists are helping to solve some of the world's most mystifying pesticide problems while uncovering new difficulties in research. Much of the important work is being carried out at the Nature Conservancy's Monks Wood Experimental Station. There, for instance, Derek A. Ratcliffe has noted a striking parallel, both geographically and in time, between the heavy use of persistent pesticides and the decline of the peregrine falcon. While this approach falls short of "scientific proof" in the strictest sense, it is extremely convincing. The population decline of the peregrine is most marked in areas of southern England where pesticide use has been heavy, and in other sprayed areas of Wales, northern England, and southern Scotland. The population decline began about 1956, exactly coinciding with the abrupt rise in the use of persistent pesticides there. Similar results have been gathered in studies made on the kestrel and the sparrow hawk.

British scientists have noted other environmental factors which have contributed to the fluctuation of bird populations in Great Britain: increased motor traffic, hedgerow destruction, overhead wires, and the reduction in numbers and variety of weed plants and their insect fauna. But research has not established definite correlations between these factors and the decline of the birds of prey, nor has any significant decline been found in either available habitat or nesting sites for the affected species.

Narrowing the area of discussion, British scientists have discovered links between pesticide residues and breeding failure. D. J. Jeffries, working with Bengalese finches, has demonstrated that there is a correlation between the intake of DDT by the female and a delay in ovulation.

"It is possible," Jeffries writes, "that organochlorine pesticides are responsible for the late breeding of several species of birds reported in the early nineteen-sixties. This could affect a population in that large sublethal doses which double the time for egg laying could cause a double-brooded bird to be out of phase with the food supply for the late broods." (It is interesting to note that the peregrine falcon, just prior to its extirpation in the New York City region, was reported to be nesting abnormally late in the season.)

The golden eagles of Scotland have been studied intensively in recent years. In the central Highlands these eagles live chiefly on grouse. Since grouse feed on heather, which is not treated with pesticides, there is no reason to believe the golden eagle population or its reproduction would be affected; and, in fact, it has not been. But in western Scotland, where the eagles live chiefly on sheep carrion, the reproduction of eagles has been drastically reduced. Although 72 per cent of the golden eagles that were studied in western Scotland between 1939 and 1960 raised young, only 10 per cent raised young during the years 1961-63. Many of these birds did not even lay eggs and (a phenomenon seldom observed in earlier years) nearly half of the birds which did lay broke their eggs in the nest!

Again, the Monks Wood scientists have made an important distinction: the golden eagle decline in Scotland has taken place in *reproduction*, not in population. Golden eagles live for about thirty years. Because the reproductive decline has been observed only since the introduction of dieldrin into sheep dips, and because the contamination of sheep carcasses is not severe enough to kill the eagles directly, the adult population has remained relatively stable.

It is hoped that the 1966 ban on dieldrin for sheep dips came soon enough to prevent permanent damage to the golden eagles in western Scotland. (Dieldrin has been replaced by shorter-lived organic phosphate chemicals.) But Britain's insularity has crumbled in many ways during the post-war era. Not the least of these is in its ability to decide for itself what toxic chemicals it will tolerate on its own soil. Britain now receives a substantial amount of DDT in its rainfall, another questionable import from the United States. One scientist estimates that the United Kingdom receives almost four times as much pesticide in rain as the Mississippi River carries annually into the Gulf of Mexico.

"It's a warm wind, the west wind," sang John Masefield, "full of birds' cries." And a great deal else besides.

*

In the United States, more indiscriminately dosed than England with persistent pesticides, there is not comparable hope for the affected species. The peregrine falcon (or "duck hawk") has disappeared as a breeding bird from the northeastern United States. The bald eagle, our national bird, has almost ceased to breed in many of its favorite haunts along the north Atlantic Coast and on the Great Lakes. A five-year study, conducted by the National Audubon Society across the United States, documented the drop in breeding success: for instance, in 1965 only four of 53 nests surveyed in Maine hatched live young. The osprey, once a familiar sight on the Northeast coast, now is declining at the rate of 30 per cent a year. This large fish hawk experienced local declines in the past, but none of them on anything like the present scale.

A colony of ospreys in Connecticut, apparently with no change in its habitat or other factors, declined from 200 pairs in 1938 to 12 pairs in 1965. A normal osprey colony is said to produce about 2.5 young per nest. A Maryland colony, with DDT residues in its eggs amounting to 3.0 parts per million, produces 1.1 young per nest. The Connecticut colony, carrying 5.1 parts per million DDT in its eggs, produces 0.5 young per nest. In most cases, the eggs simply do not hatch. DDT residues in the fish eaten by the Connecticut ospreys are five to ten times higher than those eaten by Maryland ospreys. Scientists have discovered no correlation between the ospreys' nesting success and the degree of their isolation. (Ospreys have been known to coexist with human beings in the past, nesting on chimneys and telephone poles.) There have been no food shortages for the birds. At present, the U.S. Fish and Wildlife Service is taking eggs from the healthier Maryland colony, flying them to Connecticut, and returning with Connecticut eggs for the Maryland colony. The results will be carefully analyzed.

The counterpart of England's Monks Wood Experimental Station in the United States is the Patuxent Wildlife Research Center at Laurel, Maryland. The center is operated by the Bureau of Sport Fisheries and Wildlife of the U.S. Fish and Wildlife Service. Since World War II much of the research in the United States into the effects of pesticides on wildlife has taken place there. The principal work at Patuxent during recent years has been carried out by William Stickel and his wife, Lucille F. Stickel.

"They are very careful, conservative and able scientists, constrained to report what their findings are and to leave interpretation of their findings to policy makers higher in the administrative framework."

This is the evaluation of another government scientist. Occasionally, the more militant scientists outside of government are impatient with the "conservative" approach followed by the Patuxent group. It is true that much of the research that comes out of Patuxent is "raw." Outside scientists, anxious to draw on the work completed by the Stickels often must reduce it to usable or quotable form. At times the Stickels' refusal to draw conclusions has tempted the chemical industry to distort their findings, implying that pesticide effects are nil.

However, the Stickels and their colleagues at Patuxent have worked for many years under difficult conditions. Appropriations for the valuable work at Patuxent must come from Congress. Too frequently, congressmen appropriate funds only for those projects which will produce an immediate tangible result. During the 1950's they thought research on pesticide effects worth only $52,000 a year. That sum has risen to $1.5 million a year, but it is still far too small, and might be slashed at any time. In other words, congressmen will grant money for research directed at pests (for there lie agricultural profits) but not for research directed at the effects of pesticides.

Among some congressmen dependent on agricultural interests, there is hostility toward research which may blemish the record of pesticides. Mississippi's Congressman Jamie F. Whitten is among those who seem to feel that appropriations normally funneled to Patuxent would be used to better advantage by the U.S. Department of Agriculture. Officials at USDA fan these flames by claiming that the Patuxent researchers leak information to conservationists about the hazards of certain pesticides to wildlife.

Under these circumstances, the Patuxent group has had to move cautiously. It simply locates the problems and examines them. Lucille Stickel has been especially involved in the most ticklish pesticide problems since the introduction of DDT; recognition of her many services came in 1968 when President Lyndon B. Johnson presented her with a Federal Women's Award.

"The role of Lucille Stickel is such that she is *the* voice on pesticides from within the Department of the Interior," says a scientist at a leading university, "and when anybody anywhere in government wants to know something about pesticides and wildlife, he turns to Interior. Thus Lucille Stickel is *the* voice of the U.S. Government on the subject."

This scientist is aware of the difficulty of the Stickels' position, and impatient of the system that shackles them.

"Lucille sits there with USDA breathing down her neck, and all sorts of other pressures on her, probably including a cut in funds if she gets out of line and makes an 'irresponsible' remark.

Patuxent is subject to all sorts of political pressures that we are not. But I do feel that the ultra-cautious attitude of Patuxent has allowed toleration of certain practices that should long ago have been eliminated. For instance, everyone at Patuxent knows exactly what happens when DDT is sprayed on elms, and dieldrin is applied at two or three pounds per acre for Japanese beetles. This is not a controversial area. It is a 100 percent fact area — *proven.* Yet these practices go on. Here is where, it seems to me, Patuxent should have stepped forward and squawked as loudly as possible."

"One of the most important tasks of biologists is to relate environmental pollution to the welfare of man," says Eugene H. Dustman, Director of the Patuxent Wildlife Research Center. "It would be disastrous if our programs were cut off now. I hope the politicians don't sell us down the drain."

One of the programs Dustman refers to, and to which scientists on both sides of the Atlantic look with the keenest interest, is Patuxent's sparrow hawk research. (The American sparrow hawk, like the peregrine and the English or European kestrel, is a falcon; the English sparrow hawk, an accipiter, belongs to an entirely different genus of hawks.) The American sparrow hawk (*Falco sparverius*) is not an endangered species. But the Patuxent biologists, alarmed by the decline of the bald eagle and the peregrine falcon, needed a "guinea pig" which was easier to handle than the eagle and more abundant in the wild than the peregrine.

Accordingly, young sparrow hawks were captured in the wild and used to build a captive population. These birds were known to nest in boxes, and proved to breed readily in captivity. Each pair was established in a long, outdoor cage containing a nesting box and plenty of ground cover. In 1966 the birds were divided into three groups and a long-term feeding study, involving both DDT and dieldrin, was begun.

One group of birds was fed one part per million dieldrin and five parts per million DDT in its diet, a dosage approximately equal to what might be expected in the wild; a second group was given three parts per million dieldrin and 15 parts per million DDT; the third group, the "controls," received no pesticides. The results disclosed poorer nesting success for the two groups being fed pesticides, while some of the "high-dosage" males died (so that there was a drop in population as well as in reproduction). The experiments with sparrow hawks (as well as similar ones with mallard ducks) also confirmed, under laboratory conditions, some remarkable findings made by scientists in the field on both sides of the Atlantic. A cause-and-effect relationship was established at Patuxent between sub-lethal pesticide poisoning and the thickness of eggshells.

*

This brings us to another fascinating point in the increasing convergence of evidence against the chlorinated hydrocarbon pesticides. The recent research has focused on calcium metabolism. There has been speculation that birds are such profitable objects of pesticide research because they contain a higher level of lipids (or fatty substances) in their plasma, especially at egg-laying time. DDT, extremely fat-soluble, would tend to concentrate more readily in birds and their eggs than in mammals.

A number of the affected species of birds, especially the birds of prey, share this characteristic: the birds destroy their own eggs. Some ornithologists have noted an unprecedented restlessness among these birds at their nests. Birds have been seen to leave their nests frequently, exposing their eggs to the rigors of heat and cold; to leave the nest abruptly, breaking their eggs; and to peck their eggs to pieces.

Ratcliffe in England had been especially interested in this matter of eggshells. In one sense, it is fortunate that Englishmen in the past have been ardent oölogists, or egg collectors. Ratcliffe and his colleagues have measured the thickness of eggshells kept in museum collections since before the turn of the century. Thicknesses remained constant for over fifty years, until just after World War II, when DDT was introduced to the environment. Suddenly a sharp dip in both the thickness and the calcium carbonate content of the shells appeared in the eggs of such birds as the peregrine falcon, the sparrow hawk, and the golden eagle. The decreases range as high as 25 per cent, and there has been no recovery.

Biologists note a remarkably similar weakening of eggshells among bald eagles, ospreys, and peregrine falcons, all declining species, in the United States. Recent examinations of specimens collected in California by museums during the years 1947-53 reveal that the weight and thickness of peregrine falcon eggs showed a decrease corresponding almost exactly with those measured by Ratcliffe in England. Massachusetts peregrines also began to be affected in 1947, shortly after the widespread introduction of DDT. Significantly, the American red-tailed hawks, golden eagles, and great horned owls, whose shells apparently have not been affected, show no population decline that cannot be explained by shrinking habitat.

"When one picks up an egg that has been thinned by pesticidal doses," Patuxent's William Stickel says, "the egg is cracked by the touch of the fingers."

In measuring eggshells for scientific purposes biologists use sensitive micrometers developed by the poultry industry. Such measurements have helped to add another point to the indictment of DDT: in 1967 biologists collected herring gull eggs from five states; those containing the highest residue levels of DDT and its toxic breakdown product, DDE, proved to be thinner than those with lower residues.

The discovery that eggshells are growing thinner has led to an increasing interest in the effects of the chlorinated hydrocarbon pesticides on calcium metabolism. It has been found, for instance, that the growth of oyster shells, which are more than 90 per cent calcium carbonate, was inhibited by concentrations of DDT in the water as low as one part per *billion*. Dogs fed on DDT developed tremors and died. Yet if DDT-poisoned dogs were injected with calcium gluconate (even after they had lapsed into tremors) they quickly recovered.

These and similar discoveries suggest a possible explanation of the mechanism by which DDT acts upon its victims: DDT, and perhaps other chlorinated hydrocarbons, interfere with normal calcium metabolism. This sort of interference can affect organisms in two ways: by killing them outright, or by attacking their reproductive capacity. It is the second of these effects that eventually might prove the most serious for wild populations.

In its acute action, DDT attacks the nerve axon, competing with calcium ions for sites on the axon's surface and causing axon decalcification. This disruption leads in turn to hyperactivity of the nerve, repetitive firing, tremors and death. Symptoms of this kind often have been observed in organisms deficient in calcium concentrations.

"Presumably the mechanism of acute toxicity operates to kill adult birds when lethal concentrations in the brain are reached," writes Stony Brook's Charles F. Wurster, Jr. "This can include large carnivores like hawks and eagles. The involvement of calcium in this mechanism would explain why bird kills usually include more males than females, with males showing greater susceptibility and females succumbing only after accumulating higher residue concentrations in their tissues. Females would be more resistant than males because estrogen raises their blood calcium, thereby bathing nervous tissues in a higher concentration of calcium and affording females a degree of protection not shared by the males."

The mechanism by which DDT in sublethal amounts may affect a bird's ability to reproduce is through the increased production (or induction) of liver enzymes which metabolize steroid sex hormones. This process leads to a breakdown in estrogen. Since estrogen helps to mobilize calcium in a bird's body, one result, among others, is the production of eggshells deficient in calcium. Thinner, more fragile eggshells become the norm.

In California, brown pelicans now crush their eggs simply by sitting on them in the nest. Shells found recently in eagle nests are not thick enough to protect the embryo; one of these eggs had no shell at all!

Furthermore, organisms affected by calcium deficiencies under other circumstances have been observed to behave in a restless or irritable manner, or to crave calcium in their diet. Again, we have a plausible explanation for both the thinner eggshells, and the odd behavior of the adult birds which smash and even eat their own eggs.

*

Writing in *Science* late in 1968, Joseph J. Hickey and Daniel W. Anderson, wildlife ecologists at the University of Wisconsin, made a dramatic statement about their recent work in the field (as well as that at Patuxent) on avian eggs.

"We have reached these conclusions," the authors wrote. "(i) Many of the recent and spectacular raptor population crashes in both the United States and Western Europe have had a common physiological basis; (ii) eggshell breakage has been widespread but largely overlooked in North America; (iii) significant decreases in shell thickness and weight are characteristic of the unprecedented reproductive failures of raptor populations in certain parts of the United States; (iv) the onset of the calcium change 1 year after the introduction of chlorinated hydrocarbons into general usage was not a random circumstance; and (v) these persisting compounds are having a serious insidious effect on certain species of birds at the tops of contaminated ecosystems."

10. The Human Toll

EARLY IN THE MORNING of June 3, 1967, an eleven-year-old boy in Doha, the capital city of the shiekdom of Qatar, fell to the ground unconscious. Before he could be taken to the Government Hospital he lapsed into convulsions. Hardly had the doctors begun to treat him with phenobarbitol and other drugs than a stream of patients, similarly afflicted, were carried into the hospital. Many of them suffered as well from nausea, vomiting, mental confusion, and severe abdominal pains. Almost 500 people entered the hospital; seven of them died.

Clearly, this Persian Gulf city of some 40,000 souls had been struck by a major disaster. At first the city's water supplies were suspected of carrying disease. Analysis quickly forced doctors to discard that notion. Then, as many of the victims showed a rather rapid recovery, questions were asked. A pattern began to emerge: almost all of the patients had been stricken soon after breakfast, and all had eaten bread. Furthermore, the victims had obtained their bread at the same bakery — and the baker and three of his employees were among the victims. The case seemed to be closed when the authorities learned that the baker often used a benzene hexachloride insecticide to control insects, mice, and rats about his premises.

There was no further trouble that month. Then, on July 2, thirteen persons were taken to the hospital suffering from nausea and convulsions identical with those observed the month before. All recovered. The authorities learned that they were members of a family which operated a bakery. On that day they had not opened for business, but had baked only enough bread for their own consumption. There were several mystifying developments here: the owner never used insecticides about his bakery, and the flour was the same brand used by the baker implicated in the earlier poisoning.

The next day another blow fell. Almost 200 people, displaying the familiar symptoms, were brought to the Government Hospital. Seventeen of them died. All had eaten bread baked at a third bakery in a different section of the city, but the same brand of flour was noted here as before. Tissue from the stomach of one of the victims and samples of the suspected bread and flour were flown to laboratories in England and the United States. The contaminating substance was definitely identified as endrin.

The local authorities, assisted by detectives sent from New Scotland Yard, set out to track down the source of the endrin. At first the investigators seemed to come up against a stone wall. Endrin had never been used, or even imported, at Doha. The suspected flour, it was learned, had arrived at Um Said, Doha's port, late in May aboard a freighter from Houston, Texas. It had been stowed in cloth sacks on a lower deck. When the investigators studied the ship's cargo-stowage plan, they learned that a quantity of emulsifiable concentrate of endrin had been stored just above the flour in five-gallon pails. When the endrin had been unloaded finally, in Iran, seventeen of the pails were found to be leaking, and two completely empty.

*

The people of Doha had been the victims of a dreadful accident, distinct in its origin but really of the same nature as explosions, plane crashes, and other by-products of man's modern quest for the good life. It is part of the cost he pays on his quest. This cost in human lives may be traced to the same causes as those that lay behind the dramatic wildlife kills of the 1950's and early 1960's. Ill-conceived control programs and the careless handling of pesticides themselves killed untold numbers of fish, birds, mammals, and beneficial insects. The slaughter could be attributed in large part to the "misuse" of pesticides, persistent and nonpersistent alike. Only later did American and British scientists begin to explore the less dramatic but more insidious hazards posed to wildlife by the persistent chlorinated hydrocarbon pesticides.

What we know today about the cost of pesticide use in human health and lives is chiefly restricted to the first, dramatic phase. We hear about the accidents, frightening, but isolated in their effects. The World Health Organization (WHO) has made a strenuous effort to avert disasters of the sort that occurred at Doha by means of a series of international agreements dealing with the transport of poisonous materials. No one would be likely to transport explosives in a haphazard manner; but pesticides, equally dangerous, are treated as casually as toothpaste or deodorants.

The technical literature abounds in descriptions of appalling disasters caused by an equally appalling negligence. Concurrent with the Doha incident there occurred a case of widespread endrin poisoning in Hofuf, Saudi Arabia; again, the flour had been contaminated in the hold of a ship that also contained endrin (the flour bags were badly stained, and the print on their labels blurred by the leaking poison).

In September, 1967, seventeen children died and 600 others became ill after eating pastries and sweet rolls in Tijuana, Mexico. The sugar used in the preparation of the pastries was contaminated in a warehouse, where it had been stored next to containers of parathion. Two months later eighty persons died and 600 became ill after eating bread made from parathion-contaminated flour in Chiquinquira, Colombia. It seems that a truck carrying both the flour and the insecticide made a sharp turn, breaking several of the parathion containers and spilling the contents over the sacks of flour. No one bothered to mention the spillage to the baker who received the flour. (The U.S.

Department of Transportation now prohibits the shipment of certain classes of toxicants together with other materials.)

The negligence and ignorance with which we often handle pesticides were tragically demonstrated in Turkey during the late 1950's. Wheat seeds, treated with the fungicide hexachlorobenzene in anticipation of their use in planting, found their way instead into food channels. More than 3,000 children who ate the contaminated bread developed Turkish porphyria, an ailment that one experienced physician has called "a pretty awesome disease." Among the deformities observed in these Turkish children four years later were the arrest of their growth, sensitivity to light, dark pigmentation, and an overgrowth of hair which sprouted in dense tufts all over the forehead.

Children are the chief victims of pesticide accidents. The U.S. Department of Agriculture, in a survey of such accidents, reports that 65 per cent of the victims were six years old and under. In many cases, pesticides had been left about in pop bottles or other unmarked containers; illness or death followed when children drank from, or even played with, the containers.

Farm workers also are especially liable to pesticide poisoning. The chemical industry likes to point out that accidents occur simply because the worker fails to read the instructions printed on the pesticide container's label, but in many areas the work force is composed mainly of unschooled migrants. How does the worker protect himself if he is illiterate or reads only Spanish?

Many deaths occur when farmers, after wearing protective gear while spraying their fields, temporarily go without it. Dr. G. A. Reich of the U.S. Public Health Service explains why this is so.

"In South Florida, where it is hot and sticky year round and a man has to work outside, it's hard to expect a worker to wear a slicker and a respirator and boots and gloves while he is driving along on a tractor spraying pesticides. He'll like to die from the heat of his protective gear. These things have to be considered in light of the particular situation. If a man is using something like TEPP [an organophosphate] you would be inclined to encase him in cement before he starts off, so that he would be safe. But if he is using malathion, then you might want to be certain that he wore gloves, boots and a slicker while mixing it, but then perhaps only a slicker while he drove the tractor."

Dr. Clyde M. Berry of the Institute of Agricultural Medicine at the State University of Iowa is familiar with the frequent occurrence of illness among farmers who use pesticides.

"The medical literature on illnesses from farm chemicals is remarkably meager," Berry has said, "and a tempting conclusion would be that no illnesses are resulting from their use. Apparently farmers expect discomfort and minor symptoms and do not seek medical attention. Over half of the group [under study] believed that they had been adversely affected. None had seen his physician."

Studies made on the effects of pesticides on the pilots of spray planes have turned up vital information. The high incidence of crashes among crop dusters has been attributed to mild cases of pesticide poisoning; although not made seriously ill, the pilot's reflexes are dulled so that he cannot follow the precise patterns called for in aerial dusting. The lengthy exposure of these pilots to organic phosphates also has brought on mental disturbances of considerable duration. After a study of spray pilots, the Federal Aviation Agency reported that "recovery from schizophrenic and depressive symptoms requires from six to twelve months following removal from further contact with the toxic agents." The report continued:

"Chronic exposure is associated with anxiety, uneasiness, giddiness, insomnia, somnambulism, lassitude, drowsiness, tinnitus, nystagmus, dizziness, pyrexia, paralysis, paresthesias, polyneuritis, mental confusion, emotional lability, depression with weeping, schizophrenic reaction, dissociation, fugue, inability to get along with family and friends, and poor work performance."

The affected pilots were not guilty of negligence. All were professionals who kept their planes clean and their respirators and other equipment in good condition. "Of 23 subjects in one study," the FAA says, "all of whom used carbon-filter respirators and wore rubber gauntlets and coveralls, 18 had mild to moderate symptoms, four had severe symptoms but recovered, and one died."

In one recent California crash, the pilot was found to have been fatally poisoned by his own spray, which consisted of TEPP and phosdrin. The accident and fatality rate for aerial spray pilots is second only to that of pleasure craft, where drinking plays an important role. The Public Health Service now is making a careful investigation of all fatal spray-plane crashes.

"We've found that for two successive years 11 states have been responsible for 69 percent of all fatal plane crashes," Dr. G. A. Reich of PHS says. "These states are primarily southern and western. Considering causation of the crash and contributory factors, the pilot factor is involved in 75 percent of the crashes, and the pilot factor is the sole cause in better than 30 percent of the crashes. So we feel that this is going to be a very valuable investigation. We hope to determine what role the pilot's exposure to pesticides plays in these crashes."

A recent study in California showed that spray plane pilots on their way home from work became involved in an abnormally high number of auto accidents. The deadly cargoes carried by crop dusters also increase the hazard in other ways both to the pilot and to residents near the target areas. In California's heavily farmed valleys, temperature inversions cause the most carefully directed spray patterns to drift chaotically away from the target. Scientists studying DDT have found that, *even under ideal conditions*, only about 50 per cent of aerial sprays, released at treetop level, reach the target area; the rest is probably dispersed in the atmosphere as small crystals.

Indeed, many Americans live perpetually in a sea of pesticides. Rachel Carson once spoke of "the right of the citizen to be secure in his own home against the intrusion of poisons applied by other persons." Despite this right, a Public Health Service scientist reported in 1966 that the "air over many communities is contaminated with substantial amounts of a wide variety of insecticides." The primary source of these poisons is the agricultural land around a community, as in Fort Valley, Peach County, Georgia, which has heavily sprayed peach orchards in its environs and a pesticide formulating plant nearby. DDT was found in *all* samples of air analyzed at Fort Valley. Elsewhere, scientists find that the air above fields treated with aldrin or dieldrin is toxic to flies!

What happens to these drifting chemicals? If a person is not allergic to them, the chances are that nonpersistent pesticides will not seriously harm him in the dilute quantities usually encountered in open-air situations. On the other hand, the chlorinated hydrocarbons, while quickly losing their control effectiveness, will be subject to magnification in non-target food chains.

An invasion of privacy which is almost as difficult to justify as widespread aerial spraying is the application of persistent pesticides to rugs and clothing. Most rugs that one buys today are pretreated with dieldrin or similar poisons by the manufacturer to prevent damage by carpet beetles. Recently U.S. Fish and Wildlife Service scientists investigated toxic substances in a Wisconsin river. A clothing manufacturer, they learned, mothproofed his product with dieldrin, then dumped the residues into the river nearby.

"Judging from the amounts we found," a Fish and Wildlife Service scientist said, "when you start to sweat in that fellow's suits you're liable to feel a little whirly."

Despite warnings by both the Public Health Service and the American Medical Association that they may be hazardous to human beings, USDA continues to authorize the use in restaurants and other public places of vaporizing devices which contain lindane pellets. These devices are not permitted to be sold for use in homes. According to a government report, lindane (a chlorinated hydrocarbon insecticide) "has been implicated directly or circumstantially in cases of serious bone marrow failure." Millions of these continuously operating lindane vaporizors have been sold in the last fifteen years, says AMA, and USDA's Agricultural Research Service apparently sees no reason to dampen the flourishing market. The Agricultural Research Service said it "would continue to register and re-register lindane pellets for use in continuously operating vaporizors in food handling establishments and places of work on the basis that such uses had already been accepted for other registrants of lindane pellets and that refusal to accept such uses would be discriminatory to the applicant."

The worried citizen who wants to get away from it all may escape the omnipresent chemical rain if he takes a slow boat to Cathay, but not if he chooses to go by air. Most planes on international routes today are fumigated for mosquito control with DDT or other pesticides. The risk of poisoning for the passenger is thought to be small, although there is less assurance about the effects of constant exposure of crewmen to these pesticides. An airline, which might worry about the influence of a brace of preflight martinis on its pilots' judgment and reflexes, apparently sees no cause for alarm in their pesticide intake.

The use of pesticides is notoriously abused about the house and garden. If one suburban homeowner sprays his garden and kills the predator birds and insects as well as the "pest," he automatically creates problems for everyone else in his neighborhood.

"When I give a lecture I am often asked by someone, 'But what about my *roses?* I must spray to save my *roses,*'" William H. Drury, Jr., of the Hatheway School of Conservation Education says. "I always reply with a question: 'Why do you garden?' If the homeowner says he gardens for recreation — for something to do — then I ask: 'So why are you trying to save time? Pesticides are used mainly to save time, to increase efficiency.' But if he is simply competing with his neighbor to raise prize roses, then I forget about him because that is a pretty poor motivation for gardening."

The average suburban homeowner is a victim of the chemical salesmen. Pesticides occasionally have their place in home and garden, but sales campaigns have convinced the average suburban homeowner that he must keep at hand an arsenal of chemicals sufficient to stave off an African locust assault.

"The gardening stores are full of poisons," a horticulturist writes from Oregon. "Some of them are so dangerous that they should not be sold other than on a prescription from top agricultural experts in the government, and then only to qualified and reliable users."

The use of parathion, for instance, is illegal in New York City without a special permit from the Health Department. Yet in Miami and its suburbs, where the climate is favorable to insects and fungi, peddlers go from door to door selling parathion in unmarked paper bags. PHS has investigated 211 poisonings in that area in recent years, 112 of them fatal. Most of the victims are young children, who find bags of parathion around the house. PHS investigators have no idea how many poisoning cases occur in the United States each year because the law does not require these cases to be reported to the authorities; based on their studies in Florida and Texas, investigators believe that many states are inclined to underestimate the problem because of the large number of unreported cases.

Ironically, the homeowner often wastes his money on pesticides.

"It is not necessarily harmful to a tree to have insects feeding on it, any more than it is to have a gardener pruning it," LaMont C. Cole of Cornell has written. "The suburbanite who goes to the store and buys something 'to stop the bugs from eating up my trees' is likely to be in for a rude shock. Even if he does not poison his lawn, his cat, the songbirds or himself, he may unwittingly set in motion a train of events with no predictable destination. Perhaps it will be just as well if the insects are resistant."

11. The Unknown

CLEARLY man does pay a price for using pesticides to protect his food, fiber, and ornamental plants. Despite the wide publicity given Rachel Carson's warnings seven years ago, man continues to expose himself to a broad range of poisons. Much of this exposure leads to hazards of an acute nature; the exposure can be restricted by education, common sense, and a few simple precautions. But man also submits himself to dangers whose extent and complexity he may not recognize in time. Just as the Bermuda petrel serves as a stark reminder of the danger to wildlife, the "Case of the Bolivian Cats" sounds a warning for mankind.

*

Medical World News, in its issue of July 11, 1965, told a remarkable story about man's creation of an epidemic of Bolivian hemorrhagic fever. A severe epidemic of this disease (generally called "black typhus") broke out in the upland Bolivian town of San Joaquin in 1963. More than 300 residents died. At the height of the epidemic, a team of United States epidemiologists, led by Dr. Karl M. Johnson of the National Institute of Allergy and Infectious Diseases, flew to Bolivia. They brought the disease under control, then set about trying to isolate its origin.

The viral agent was quickly identified. Various insects such as mosquitoes, mites, and ticks were examined and eliminated as possible vectors. In searching for other possible vectors, Johnson and his colleagues learned that a curious invasion of mouse-like rodents called *lauchas* (*Calomys callosus*) had invaded the town shortly before the start of the epidemic. These rodents had seldom been seen in San Joaquin's houses before that time. The epidemiologists set out traps and poison for the *lauchas*. When they disappeared, the black typhus disappeared too. Examination of the dead *lauchas* disclosed that they were the vectors of the deadly virus.

But why had these wild rodents suddenly taken to human habitations?

"Five years earlier, cats had been ubiquitous in San Joaquin," Dr. Johnson said. "But when we arrived, there were only about a dozen of the animals left in all of the town's 340 households. People told us that their cats had been dying off for the past several years, the victims of a strange 'cat disease.' This struck us as odd, because of the peculiar neurologic symptoms: the cats would have the shakes, get sick, linger on a few days, and die."

Had the cats died of black typhus too? Dr. Johnson expressed cautious doubt. The cats had begun to die long before the disease appeared among the human residents of San Joaquin. When Dr. Johnson injected cats with the virus, they showed no ill effects.

"The cats' symptoms looked to us like DDT poisoning," he said. "We knew that DDT and dieldren had been used for several years to control malaria in San Joaquin. Later, when the hemorrhagic fever broke out, more DDT powder was spread in order to kill its presumed insect vector."

Nevertheless, Dr. Johnson secured one of the dead cats which had been collected earlier by the local authorities and kept in a deep freeze. He shipped the carcass to Dr. Wayland J. Hayes, Jr., chief of the toxicology section of the United States Public Health Service at Atlanta, Georgia. Hayes' report confirmed that the concentrations of DDT in the cat's brain were consistent with a finding of death by DDT poisoning.

It was then that Dr. Johnson began to close the epidemiologic ring: San Joaquin's cats had once acted as a natural control on the neighboring *laucha* population. When DDT was applied to the walls of the houses, the cats picked it up on their fur, then ingested it while licking and preening. With their natural controls wiped out, the *lauchas* invaded the town, carrying with them the fever virus which they spread through their urine around household food and water supplies.

DDT had been brought to San Joaquin to tip the balance of nature in favor of man by eliminating the malaria-carrying mosquitoes. Three hundred townspeople paid for the program with their lives.

*

Since the publication of *Silent Spring*, many other scientists have added fuel to the notion that we may be letting ourselves in for a nasty surprise.

"It is probable that continued exposure to low levels of toxic agents will eventually result in a great variety of delayed pathological manifestations, creating physiological misery and increasing the medical load," one distinguished physician writes. "The point of importance here is that the worst pathological effects of environmental pollutants will *not* be detected at the time of exposure; indeed they may not become evident *until several decades later*. In other words, society will become adjusted to levels of pollution sufficiently low not to have an immediate nuisance value, but this apparent adaptation will eventually cause much pathological damage in the adult population and create large medical and social burdens."

If there has been a significant advance since the publication of *Silent Spring* in this area of long-range pesticide effects on human beings, it is in the expansion of the realm of the unknown; *our ignorance of the subject has broadened and deepened.*

Official statements telling us that we have nothing to worry about are no longer so assuring as they were in those days before the credibility gap engulfed government scientists. The revelations about the true state of affairs concerning radioactive fallout poisoned the public's trust. An American thalidomide tragedy was averted by one stubborn woman with her finger in the dike. More recently, the Food and Drug Administration has been accused publicly of approving inadequately tested drugs. Doctors and dentists used X-rays with great confidence for many years, but with the abrupt rise in thyroid cancer, birth defects, and other indices of danger, there has been a searching reappraisal of the X-ray machine's role in medical treatment. (No one even knows what constitutes a dangerous radiation dosage.)

And what of DDT residues on tobacco? Though residue levels of seven parts per million and less have been set by FDA for many leafy vegetables such as lettuce and spinach, levels on marketed tobacco leaves have been noted as high as 53 parts per million. Even official pesticide residues on foods are suspect. An authority on lead poisoning made the following statement before a Senate subcommittee in 1966:

"The maximum permissible level of lead in apples set by the U.S. Department of Agriculture is seven parts per million. If lead were to be added to the food of humans to the extent of seven parts per million they would die of classical lead poisoning."

And lead is one of our oldest pesticides!

Admittedly, it is as difficult for pest control experts to foresee all the unwanted effects of their chemicals as it is for doctors to foresee all the toxic, as well as the therapeutic, effects of their drugs. The agri-chemical industry argues that a poison ceases to be a poison when it is administered in small quantities (the acute toxicity of DDT approximates that of aspirin). But two points should be made here: aspirin and many other "poisons" are not cumulative in the body, as the chlorinated hydrocarbons tend to be; and we still know little about the interaction in the body of the many minute quantities of poison to which we are simultaneously exposed. Medical literature abounds in references to the toxic interaction of several drugs — and one class of drugs is known to cause a toxic reaction with cheese!

Allergies to the increasingly dense chemical fabric of our environment are quite common. Doctors are aware of acute reactions to the mixtures of chlorinated hydrocarbon pesticides and solvents used in Dutch elm disease or mosquito control. Susceptible individuals have lapsed into unconsciousness when their homes were treated with insecticides, sometimes several days after the spraying was finished. Other reactions to chemicals include bronchitis, asthma, headache, hives, and gastrointestinal disorders.

The World Health Organization, which uses pesticides extensively in its antimalarial campaigns, has voiced fears in recent years that the chlorinated hydrocarbons may be a cause of liver damage among human beings. Workers in the Soviet Union who have been occupationally exposed to DDT and other chlorinated hydrocarbon pesticides for more than ten years were found to have pronounced disturbances of the protein and sugar metabolism in their livers. This is one of the human health areas in which there is a great deal of research taking place. Some scientists actually have seen in the effects of pesticides on the liver certain favorable, rather than destructive, processes. They see, not damage, but an increase in functional activity of the liver; both malathion (an organic phosphate) and methoxychlor (a chlorinated hydrocarbon) are less toxic to human beings than many other pesticides because liver enzymes become activated to deal with them. But this remains a very cloudy area. Research on birds and rodents, as we have seen, indicates that the production of liver enzymes inactivates the steroid sex hormones. Will further research establish the existence of similar effects in human beings?

The specter of cancer has hung about pesticide use ever since *Silent Spring* was published. The painfully slow progress made by scientists in their efforts to understand and find cures for the various cancers goes on also in the attempt to establish a link between cancer and pesticides. The links are no more solid today than they were at the time that Rachel Carson posed her searching question: are the synthetic organic pesticides carcinogenic? *The scientific community is not ready to commit itself to an answer.*

But every reputable physician and scientist advises extreme caution in this matter. They do not go along with the chemical industry's contention that *no* evidence means *negative* evidence. In 1967 the New York Academy of Sciences sponsored a conference on the subject, "Biological Effects of Pesticides in Mammalian Systems." Scientists from the United States and Great Britain displayed deep concern that the mutations observed in animals exposed to pesticides might also occur in humans. Bits and pieces of circumstantial evidence (reminiscent of the careful construction of the case against cigarettes) were reported at the conference. British scientists maintained that the breakdown products of all the major categories of pesticides were capable of doubling the mutation rate in man. (Mutations are changes in the genetic material which usually lead to profound and harmful consequences.) Other scientists revealed that fungicides such as captan and folpet — both considered "safe" by the authorities — may be capable of causing mutations in man. Pregnant women were warned about spraying their roses with these fungicides.

Ironically, the evidence of chemical pesticides' destructive effects on wildlife is now so conclusive that scientists prefer to dwell on it, rather than on the human side of the picture, in their public testimony. Only occasionally are the human aspects brought up. One of these occasions was the hearing at Madison, Wisconsin, in 1969 to deliberate on the prohibition of DDT in that state. Richard Welch, a pharmacologist with Burroughs Wellcome and Company, which manufactures drugs, told the hearing that the sex hormones affected in rats by certain DDT-activated enzymes are the same ones found in man. In fact, the amounts of DDT now stored in human fat approximate the amounts needed to affect rats.

"Thus, if one can extrapolate from animals to man," Welch said, "then one would say that the changes in these enzymes probably do occur in man."

Robert Risebrough, a prominent biologist at the University of California, alluded to the same effects of DDT and the other persistent pesticides. He pointed out that the Food and Drug Administration "hasn't taken any consideration of the enzyme inducing capacity of these substances. This is a decision which the FDA will have to make sometime in the near future. Will it permit an increase in the activity of these enzymes in our liver? No responsible person could now get up here and say that this constant nibbling away at our steroids [or sex hormones] is without any physiological effect."

A number of chemical pesticides produce cancers in mice. Dr. James T. Grace, director of the Roswell Park Memorial Institute, told a subcommittee of the New York State Legislature in 1969 that his institute had compiled a great deal of evidence about this. Like most other scientists, he is reluctant to extrapolate the results of experiments with animals to human beings. (It might be noted, however, that in conducting their research into the "safety" of pesticides, the chemical companies do just that.) Grace concluded his testimony with this warning: "If we find these chemicals create problems in lower forms, then we must be extremely careful on how we gamble on their use in our environment."

The National Cancer Institute is engaged in studies which already indicate that pesticides act as carcinogens in mice. Commenting on these discoveries to *Medical World News* in 1969, Dr. Malcolm M. Hargraves of the Mayo Clinic said: "Since the advent of pesticides in 1947, I've seen and taken inquisitive personal histories on 1,200 cases of blood dyscrasias and lymphoid diseases. Every patient at some time or another had great exposure to a pesticide, an herbicide, a paint thinner, a cleaning agent, or the like. And I wouldn't exempt organophosphates, which the National Cancer Institute isn't testing at all. I've had several cases demonstrating their involvement in blood marrow depression, sudden prothrombin changes, liver insufficiency, and thrombocytopenia purpura."

Dr. William C. Heuper, a former chief of the environmental cancer section of NCI, expressed his concern to the same reporter. Pointing to studies at Miami University which showed that an unusually high rate of terminal cancer patients were found to have high concentrations of pesticide residues in their liver, brain, and adipose tissues, he commented: "That's the kind of study that gives you leads."

In recent years, gatherings of geneticists at the Oak Ridge National Laboratory in Tennessee and the Jackson Laboratory in Maine have discussed what several of their members have called "our number one health problem." The threat of genetic damage is denied by no prominent geneticist, although these scientists do not always agree on how much evidence is necessary before a suspected chemical agent must be removed from circulation. In any final decision, two haunting facts must be kept in mind:

1) *The genetic material is structurally identical in all organisms* — mice, men, and microbes — so that chemicals which damage or destroy this material in one species must at least be suspected of damaging it in other species. Scientists do not yet know all of the conditions under which this damage will *not* occur.

2) *Mutagens occur in the germ plasm before conception.* They may lie undetected through several generations, then suddenly appear in ways destructive to those descendants of the organism in which they first occurred. The mutagens may finally appear as genetic illnesses, or as a general loss of vitality.

It would be ironic if, long after the fathers restricted the use of hazardous chemicals, their original blunders would be visited on their children. One geneticist has made an estimate with which his colleagues agree: "Damage sown in the germ plasm is far more dangerous to the human race than immediate clinical complications like cancer or thalidomide, which cripple or kill a single person but are not reproduced."

One of the profoundest shocks we have experienced during this chemical age is the realization that the womb is not the secure little niche we had imagined it to be. Pesticide residues have been detected in unborn babies and in mothers' milk. DDT levels in human milk in the United States range from .15 to .25 parts per million. This means that American babies consume four times the maximum daily intake recommended by the UN, or five times the DDT content allowed in the interstate shipment of cows' milk. (British breast-fed babies ingest DDT in amounts only 70 per cent higher than the recommended limit, but, like American babies, ingest *ten times* the recommended limit of dieldrin.) Doctors for some time now have warned women not to spray their rooms with pesticides during pregnancies. Teratology — the study of birth defects — is concentrating increasingly on the role chemicals play in the tragedy that seven per cent of all babies born alive today suffer from some defect.

This, too, is an area in which our ignorance seems to expand rather than contract. So many facts are missing. No one knows, for instance, whether birth defects actually are increasing in this country. No one even knows the exact time at which ovulation takes place in women; pregnancies are not confirmed until at least nineteen days after conception, and it is during this period that many birth defects are determined. The most basic guidelines are lacking. Birth defects have been known to occur in clusters, like cases of hepatitis. In recent years "epidemics" of babies born with spina bifida (open spine) are known to have occurred in Georgia and Vermont. In 1965 four sets of Siamese twins were born in New York City within a few months. Since the discovery of this sort of "epidemic" usually comes by accident, no one knows exactly how common they are.

Only occasionally are the links established. The cause of severe mental and motor retardation among some Japanese children was discovered to be inorganic mercury, flushed from a vinyl-producing plant nearby. Fish, accumulating the mercury, apparently were unaffected. But pregnant women accumulated it too, and in crossing over the placenta the substance attacked the babies they carried.

It would be dishonest today to brand chemical pesticides as a major cause of cancer and birth defects. But it would be foolhardy to suggest that we already know the full cost to our health that we must pay for their unrestricted use in the environment.

12. In the Scales

SCIENTISTS from all over the world have contributed evidence which suggests that our present pesticide policies are disastrous to man and his environment. If we are paying a price for the manner in which we use chemical pesticides, we might rationally assume that the benefits to be had clearly outweigh the cost. But have the benefits been assessed? And have these benefits been weighed against the cost?

Since pesticides are classed as "economic poisons," their use generally is justified on the ground that they are profitable. (A similar odor of profit hung about poisons during the time of Louis XIV when they were referred to, somewhat sinisterly, as "inheritance powders.") Pesticides certainly are profitable to the chemical industry and its outlets. Sales of chemical pesticides in the United States rose 10 per cent over the previous year in 1965, then jumped an astonishing 18 per cent in 1966. Manufacturers gross over $600,000,000 today, and retail outlets more than double that figure.

The trend in pesticide use is upward at progressively increasing rates. This is what the U.S. Department of Agriculture's *Pesticide Review* said in 1967: "Expanding worldwide food production generates a demand for effective means of pest control and encourages a rapidly growing pesticide industry. In the United States farmers, industrial and governmental users, and homeowners all are increasingly aware of the usefulness of chemicals in pest control. Export markets show a rising demand for pesticides which is unprecedented."

Industry, then, cheerfully justifies the widespread use of pesticides by the cost-benefit ratio. But is this ratio always valid? Apparently not, because the cost (or risk) has not been fully revealed to the public. Instead, in the absence of documented facts, the public (which should make a final decision based on social good) has been lulled into inattention by industry-sponsored propaganda.

When government and industry publications speak of the vast economic benefits derived from pesticides in general, what are the economic facts on which this estimate is based? Surprisingly, the facts are both meager and fuzzy. This was made clear recently by one of those instances of literary symbiosis that contribute needed light to a neglected subject. One element here was the publication in 1967 of a book, under the sponsorship of Resources for the Future, Inc., entitled *The Pesticide Problem: An Economic Approach to Public Policy*, by J. C. Headley and J. N. Lewis. This book in turn served as a sounding board for an exhaustive review of pesticide's costs and benefits written by the noted conservationist, Roland C. Clement, for *Natural Resources Journal* early in 1968.

The authors of *The Pesticide Problem* reflect the laissez faire concept which affirms that the sole criterion for turning any technological innovation loose in the environment is private gain. They see the problem as "one of maintaining an environment in which the goals of society tend to be achieved" (*i.e.* high levels of production). The pesticide problem, they say, exists only when "spillover effects are of such magnitude and distribution that the market place cannot adjust the values."

Ironically, the authors have made a contribution to solving the pesticide problem by their abortive attempt to establish the economic facts of pesticide use. As Clement points out in his review, they "found so little economic evidence to sum up that they had to concentrate on constructing a conceptual framework for future data gathering and the testing of alternatives." For instance, the available information on the amount of pesticides used annually is so scanty that the authors resort to figures on "domestic disappearance" of these chemicals. Nor are the positive benefits of increments in farm and forest production documented. On the other hand, the levels to which the cost of pesticide use has risen are dramatically documented: the cost of buying pesticides and applying them to cotton fields averages $12.00 a bale!

Clement observes that only by examining the pesticide problem from a specialized point of view can the authors define the problem as largely economic. He redefines it as "largely economic but inescapably a result of our generation's lack of ecological sophistication in dealing with its so-called pest problem." The farmer who uses pesticides is not a breed apart; he is as dependent as the rest of us on a "clean environment and a viable ecosystem."

The authors and their critic combine to reveal the limited work that has been done to establish the cost of chemical pesticide use in relation to its benefits. No one has calculated the damage inflicted by the pesticide user on the person, crops, or livestock of his neighbor. Many of the advances in agriculture that have been credited to pesticides really spring from the use of improved seed, fertilizers, machinery, and management. Finally, from whatever economic gain has accrued to the nation in agricultural efficiency must be subtracted the cost for what Clement calls "the socio-political and economic effects of labor displacement, as measured by the increasing public welfare burden in urban centers."

*

In discussing pesticides, conservationists rightly are concerned with ecological considerations. But they err badly if they drop the serious insect pest from these considerations. The insect (if it is a genuine pest) must be controlled in the most efficient manner possible at this stage of technology.

Rachel Carson was the first to point out to a wide audience

that pest control as it often was practiced in America did not necessarily serve the public good. Because of the nature of the persistent pesticides and the abandon with which they were applied, the local benefits gained from their use failed to offset their attendant ills. Problems arose and multiplied in a simplified environment. One scientist has compared the use of persistent pesticides in the environment to administering a long-lasting antimitotic drug to a cancer patient; such a drug might knock out intestinal epithelium and white cells as well as the cancer cells, leaving the patient vulnerable to other ills.

The arguments for and against pesticide use are clouded by several considerations. One is that pesticides (their characteristics as well as their uses) are immensely varied. Another is that their "nature" often is appraised in the eye of the beholder; a man with a financial interest in a certain pesticide, even DDT, is likely to assess its value in terms other than those adopted by a man who prizes wildlife.

The difficulty of establishing a cost-benefit ratio is clearly illustrated by the use of herbicides. Incontestably, they are an important labor-saving device. Herbicides attack and destroy unwanted plants either by direct poisoning or by accelerating their growth rates. They are not persistent but are broken down rather quickly in the soil by bacterial action.

Herbicides often have been used stupidly. In the hands of negligent crews controlling brush along roads and utilities' rights-of-way they have created monumental eyesores. In the hands of negligent homeowners they have ravaged lawns and ornamental shrubs as well as the weeds at which they were aimed. Yet they are of considerable economic value to the farmer who uses them wisely, and of aesthetic value to the skilled gardener as he shapes his landscape.

Both government and industry publications like to tell us that herbicides, in the recommended dosages, are "harmless" to wildlife and domestic animals. It is difficult to challenge that contention if we restrict ourselves to cases of direct poisoning. However, when tests conducted by the National Cancer Institute in 1969 showed that 2,4,5-T caused cancer in rats, USDA canceled its registration for use on food crops.

Further, all reports of the "harmlessness" of herbicides to wildlife neglect the stark fact that, used indiscriminately, they sometimes wipe out the habitats on which certain wildlife depends for its existence. With admirable prudence, England's Kenneth Mellanby has summed up his judgment of herbicides in a sentence studded with qualifications: "In general I am of the opinion that today weedkillers are not a major danger to wildlife in Britain."

What of the nonpersistent insecticides? Such compounds as parathion, malathion, and carbaryl have a place in integrated pest control programs, where pesticides can be used in conjunction with sound cultural practices and some of the new biological controls.* The trouble here is that they are not confined to use within agricultural plots, as they should be, but too often are spread through the landscape in needless "nuisance abatement" programs.

It must be remembered that these organic phosphate and carbamate compounds, though not nearly as persistent as the chlorinated hydrocarbons such as DDT and its allies, are extremely toxic to both man and animals. We still do not know all of their secondary effects. Used at the wrong season, for instance, they either kill birds directly or wipe out the food supply on which birds depend. Used to kill insects in Asiatic rice paddies, they kill the fish upon which the people depend for their protein. As they become more widely used, these chemicals will encounter resistance among pest insects too.

In California the lygus bug has been considered for many years a major pest on cotton. Recent experiments there suggest that the millions of dollars spent to control this insect may have been wasted. Fields treated experimentally with pesticides for the lygus bug did not yield significantly larger crops than un-

* To be discussed in Chapter 24.

treated fields; the insects apparently were harvesting the surplus buds and bolls which never ripen. Moreover, the powerful sprays used to kill the lygus bugs were also killing ladybugs, pirate bugs, and other insects which eat the eggs and larva of the more destructive bollworm.

The attempt to measure the benefits of even the persistent pesticides against their cost is frustrated in a number of ways. All nations cannot use the same yardstick. At this stage of their development it would be unrealistic to expect the hungry nations to reject foods containing what we believe to be excessive (though not fatal) residues or to become alarmed by the disappearance of certain wild creatures.

While the use of DDT is now restricted in the United States, exports continue to rise because poor nations find its price is favorable. The Agency for International Development (AID) finances pesticide purchases, averaging $20 million a year, as a part of its program; DDT comprises over 60 per cent of these purchases. Some of the exported DDT, however, is used in the World Health Organization's antimalarial campaigns. In those areas where the disease vectors have not acquired resistance to DDT, WHO still considers it the safest of all the chemical pesticides for *indoor* applications. A recent WHO committee concluded:

"The concern that has been expressed in recent years about contamination of the environment by this very stable and persistent insecticide should not, in the opinion of the committee, be considered sufficient reason for substituting other insecticides for indoor residual spraying against mosquitoes. The safety record of DDT remains outstanding."

Nevertheless, the outdoor spraying of these persistent pesticides increases throughout much of the underdeveloped world. These nations invariably lie in the tropics, where undesirable pesticide effects are likely to be magnified because of the absence of a dormant season. There is increasing concern among ornithologists in the United States for the familiar warblers and other insect-eating birds which breed in the temperate zone but winter in the tropics. And sometimes the new pesticides only increase the local people's problems. In parts of Peru and Malaysia, persistent pesticides killed all the predator insects; when the pest species produced resistant strains, crop losses became so heavy that the new pesticides had to be abandoned.

Though, as we have seen, the ring of evidence against DDT and the other persistent pesticides approaches full circle, not all of the pieces have been fitted into place yet. Thus, foot-dragging is still common. Just as the tobacco industry has fought the mounting circumstantial evidence that indicts its product as a major cause of lung diseases, the chemical industry's defenders have seized on every loose end in the pesticide problem to help them fight a delaying action against the eventual prohibition of the chlorinated hydrocarbons. Ospreys in Alaska, though carrying DDT residues equal to those found in many ospreys in the Northeast, seem to be thriving (though the research has been confined to small numbers of Alaskan birds). In England, herons seem to be holding their own, despite rather high residue levels. Some species, and perhaps races within species, are more resistant than others to pesticide poisoning.

Typical of the scientists who have shrugged off the pesticide menace is J. Robinson of the Shell Chemical Company's Tunstall Laboratory in Kent, England. Robinson has been called, by several scientists who view the chlorinated hydrocarbons with alarm, "the best of the opposition." Like a good criminal lawyer, he tries to cast doubt on key points in the prosecutor's case.

Robinson has concentrated on pointing up the disparity in residue concentrations found in stricken and healthy birds, of the same and of different species. Pointing out that more than eighty species of birds have become extinct since the seventeenth century, Robinson has looked over the present situation and found it not at all unfavorable. He attacked an important English document, "The Fifth Report of the Joint Committee of the British Trust for Ornithology and the Royal Society for the Protection of Birds," for certain statements the Report, in fact, did not include. Robinson was taken to task in a succeeding issue of the "New Statesman" by Stanley Cramp of the RSPB, who scored his utter lack of ecological considerations and his "misunderstanding" of the one document he had attempted to examine in detail.

Charles F. Wurster, Jr., is one of the scientists who has rebutted Robinson's emphasis on the apparently contradictory residue levels found in both stricken and healthy birds. Wurster pointed out that the whole laboratory testing concept of LD_{50} is based on the fact that one half the test population is killed by a certain dosage.

"The same concept applies to disease," Wurster says. "After the same exposure some will get the disease that could be caused by exposure and some will not. The same reasoning holds for Dr. Robinson's analysis of birds' eggs. He pointed out that birds' eggs from successful nests showed no less residue level than from unsuccessful nests, and therefore concluded that there was no causative relationship. But this is not necessarily true. There need not be a greater residue level in the victims than in the survivors, although there is a toxic effect. On the other hand, the increased levels found in some of the diseased cases need not show that DDT contributed to the disease."

Much of Robinson's defense of the *status quo* in pesticide use is based on the number of British birds which recently have shown no signs of a decline. Most of these birds, however, are seed-eating species which suffered mass kills before the voluntary ban on the offending seed dressings in spring sowing. The birds have, since the ban, returned to normal numbers.

Replying to Robinson during a recent conference on pesticides, J. Newman, another British scientist, had this to say: "If you see some smoke pouring out of a hotel door you do not ask for mathematical proof that the building is on fire before doing something about it. Practical decisions by those controlling pesticides have to be made on the evidence available."

*

The nature of that evidence indicates that mankind must move swiftly. Scientists, probing the extent to which the world's environment has been contaminated by persistent pesticides, have shed light on what this contamination eventually might mean if allowed to go unchecked. George M. Woodwell, a biologist at the Brookhaven National Laboratory on Long Island and an authority on radioactive contamination, is among the worried scientists.

Woodwell believes that man's dependence on the illusion of "dilution" has thrust the persistent pesticides into the forefront as our most menacing world contaminant. Man has constantly got himself into trouble in this century because of his belief that whatever pollutant he lets escape into soil, air, or water will shortly be diluted to harmlessness by the enormity of the environment. One after the other — soil, air, and water — finally have buckled under the load. Pollution has become a major problem; in most cases it has not become a disaster because the concentration of the offending matter is confined to certain localized areas — the topsoil of agricultural lands, the air over cities, or the water of individual lakes and rivers. Some contaminants finally are broken down, or leak into vaster environmental areas as new contaminants are poured in.

But it is the nature of DDT and its chemical relatives to break down very slowly. Years may pass before they disappear completely from biological systems. Though the production of DDT is dropping, this drop is offset by the rise in production of the other chlorinated hydrocarbon pesticides such as aldrin, dieldrin and endrin.

"Thus we can expect far greater changes in the world's biota than we've seen so far if we continue using these long-lived pesticides," Woodwell writes.

The most recent evidence is clear that the risk of using DDT and its chemical relatives in the open environment outweighs whatever benefits they bring to mankind.

Part III

The Flow of Knowledge

13. The Muddied Sources

"I SUBMIT that the campaign of false fear against the use of modern pesticides has, is, and will cause deaths and sufferings greater than those of World War II. It has been over 12 years since a major new insecticide has been brought to market and this is due to unnecessary controversy. During this interim, daily deaths due to starvation and malnutrition have risen from 6000-7000 per day to over 12,000 per day, not to mention the millions who have died from vector-borne diseases. These lives could have been saved had the efforts devoted to controversy been used to encourage the discovery and wider use of insect controls. Each person who has played a part in the campaign of fear must accept responsibility for his share of the unnecessary toll of human life, a toll that will continue and increase because we are still handicapped by an environment polluted by that false campaign."

These were the opening sentences of one of the most extraordinary articles to appear in a reputable scientific journal for some years. They appeared under the title, "Pesticides and the Environment," in the September, 1967, issue of *BioScience*, the journal of the American Institute of Biological Sciences (AIBS). Their author was Louis A. McLean of the Velsicol Chemical Corporation, whose pen had been raised five years before in that sabre-rattling letter (already quoted) which attempted to persuade Rachel Carson's publishers to scrap her book.

After taking the customary swipe at Rachel Carson (with a

bow to the "colorful language with which she was gifted") McLean proceeded in his article to suggest that the chemical industry publish a list of the "anti-pesticide leaders," and add to their names "the number of other variant views they have expressed." A person who questions the more widespread use of pesticides, McLean contended, also can be expected to oppose fluoridation of water supplies, food additives, public health programs, "medicine, science and the business community." Besides, he wrote, "they are actually preoccupied with the subject of sexual potency to such an extent that sex is never a subject of jest."

The response from readers and AIBS members was swift and lively. Some wrote to point out the distortions in the article, others to chide the editors for giving such a prominent position in the magazine to what they assumed to be a rather overboiled spoof. One reader replied good-humoredly to that "hilarious monograph," while another indignantly objected:

"I do not oppose the use of chemical fertilizers, fluoride in water supplies, or jokes about sex. But I do oppose the publication of irrelevancies or worse in a journal serving as the organ of a scientific professional society of which I am a member."

Apparently the editors, momentarily overcome by a rush of democratic spirit, had accepted the article to balance their publication several months before of a thoughtful and cautionary article on the abuse of our natural resources. Fortunately, the journal's readership is professional and informed; the article was treated in scientific circles with the levity it deserved. The general reader, however, no matter how well informed in other areas, usually does not have the background facts at hand to deal with the flow of unsound pesticide information. Why are the facts handed down to the public distorted or inadequate?

"When everything has its price, and more than price, and anyone is venal, what a thing is the interested mind with the disinterested motive." Marianne Moore wrote those lines in another context, but they might be applied with justification to Rachel

Carson. In swerving from the treatment of themes that were more congenial to her, Rachel Carson reviewed a large body of scientific data dealing with pesticide use and presented it in digestible form to the public. She was able to make sense of her material because she infused it with what another scientist has called "the ecological conscience."

Unfortunately, much of the literature ground out by pesticide experts lacks this broadening infusion. Too many scientists fail to see that there is no such thing as an "objective" book on an issue of social importance; an objective book about pesticides and their effects on the environment would be merely a descriptive bibliography.

If the pesticide muddle is chiefly the result of a breakdown in communications, the bulk of the guilt must lie with the academic community. Here the problem is basically philosophical. Modern techno-science has been dominated by positivism: evidence may be accepted only if it is detected by the senses. Scientists and technologists of this persuasion work by a simple one-to-one relationship which is foolproof in the laboratory for creating technological marvels and verifying certain facts.

But this sort of approach often breaks down outside in the enormously technical modern world where the creations of science and technology function on a higher level of interrelationships. The major problems of environmental pollution were not predicted in the laboratory. Bound in the straitjacket of positivism, many scientists refuse to acknowledge the warnings posted by circumstantial or correlative evidence, waiting instead for the "scientific proof" provided only by the establishment of cause-and-effect relationships. Meanwhile, for want of "proof," cigarettes continue to add to the incidence of respiratory diseases, radioactive fallout jeopardizes the health of our children, and persistent pesticides gnaw at the foundations of our viable ecosystem.

In the face of man's massive intervention in the functioning of the natural world, the scientific establishment simply filed the ominous facts and kept mum. These "silent scientists," as they have been called, spoke only to themselves, and rarely then. They did not choose to make known the facts that vitally concerned the public. They sneered at such techniques as "popularization," and recoiled in indignation from the suggestion that they cooperate with the mass media to put across the story that should have been told. For some scientists, Rachel Carson's sin was not only her willingness to tell what she knew, but to tell it in such a way that it was grasped by the public.

"The new duty of the scientist to inform his fellow man is the key, I think, to the humane use of the new powers of science," a prominent member of the scientific establishment said recently. "By this act, the scientist can open the momentous issues of the modern world to the judgment of humanity, and it is only this judgment which has the strength to direct the enormous power of science toward the welfare of man."

Even the natural sciences have resisted this message. Although technological devices related to their own special bodies of knowledge have brought about serious social and environmental problems, the scientists generally have remained silent. Their habitual reluctance to sally forth from the ivory tower has been abetted since World War II by the flood of industry money into the universities. The chemical industry, for instance, has slowly taken over the sponsorship of much academic research through grants to the colleges and to the National Research Council (the research arm of the National Academy of Sciences).

Criticism has come from within the academic establishment. Speaking recently of outside pressures, the Harvard paleontologist G. G. Simpson wrote: "It is increasingly true that the research capacity of academic biology is devoted to NASA, to Defense, or to industrial ends. Some university researchers who were more or less surreptitiously working for Defense with questionable ethics have recently been caught red-handed if not red-faced, but that does not end the matter, and most of this research imposed from outside is pointed to with pride rather than shame."

Most of the scientists thus employed by industry are "positivists" whose vision is further restricted by what one ecologist has called "the tendency to fractionate, intensify, and barricade the subdisciplines." They do not try to interpret their fields in social terms. The complexities of population biology find no place in their circumscribed world. It is no wonder, then, that they have failed to uncover the damaging, ecosystemic effects of the persistent pesticides. And so, since they have failed to detect anything, they insist that what they have not seen does not exist. The glass of their knowledge does not take in the many interdependent strands which make up the web of nature.

"Failure of scientists to criticize publicly, to any appreciable degree, programs many deem ill-judged often stems from analysis of the balance sheet of their own self-interest," another scientist wrote in a 1966 article for the *Saturday Review*. "On the positive side is the consideration that the long-term interest of their profession and the nation dictates that unwise expenditures not be made. If the public loses confidence in the integrity of scientists, the sequel could be calamitous for all. But this nebulous possibility does not outweigh present realities. The witness who questions the wisdom of the establishment pays a price and incurs hazards. He is diverted from his professional activities. He stirs the enmity of powerful foes. He fears that reprisals may extend beyond him and his institution. Perhaps he fears shadows, but in a day when almost all research institutions are highly dependent on federal funds, prudence seems to dictate silence."

Pressures of all sorts are exerted to maintain this silence. Robert L. Rudd, who may have been one of the earliest advocates of "free speech" at the University of California, paid the price usually exacted of academics who hit hard at the system. This prominent wildlife biologist's book, *Pesticides and the Living Landscape*, had been suggested by the officers of the Conservation foundation in 1958. Neither Rudd nor the Foundation's officers knew at the time where the study would lead but, under a grant from the Foundation, Rudd pursued the trail to some very unsettling conclusions.

"This book," he wrote in the Preface, "describes some of the conflict between man and other living things, and the methods we have used to reduce this conflict. It . . . may seem to show inordinate preoccupation with real and potential hazards, with imbalanced values, and with alternative control procedures. This emphasis is intentional."

Rudd completed his manuscript even before Rachel Carson completed *Silent Spring*. It was submitted by the Conservation Foundation to a prominent commercial publisher, which turned it down as a "polemic." The manuscript then was sent to the University of Wisconsin. There it languished for over a year, while some eighteen reviewers (including the entire Entomology Department) hashed over its premises. Rudd was told later that he "holds the record for reviewers" at Wisconsin. Finally, the vice-president of the University gave it his approval and the book was published in 1964. Since then it has been reviewed, mostly very favorably, over 200 times, has been published in four countries and recently was translated into Swedish.

Meanwhile, Rudd's attempt to speak out from the confines of a land grant college had brought him only personal discomfort. He lost a promotion, and his very position with the University was threatened.

"The trouble with my own efforts is the same as with the upset following *Silent Spring*," he has said. "Challenge to a basic, well-entrenched system — far more extensive and profound than most people comprehend — is simply not done. It is particularly inacceptable from someone 'inside.' I had worked on vertebrate pest control for five years and was a member of the state Agricultural Experiment Station. I was dismissed without notice or cause given from the Experiment Station in 1964."

In 1962 a publication of the University of California called "Science Guide" referred its readers to the first installment of *Silent Spring*, then appearing in *The New Yorker*. This was a usual procedure for publicizing items of interest to the students and faculty. Immediately, however, some power behind the scene applied the pressure. The editors of the guide were informed by a telephone call that Rachel Carson was not a chemist, and that *Silent Spring* was a controversial matter which had better not be discussed in a bulletin distributed by the university. Succeeding installments of *Silent Spring* were not mentioned in the guide.

Much more recently, and the width of a continent away, certain members of the Maine State Biologists' Association felt similar pressure from their superiors. At the end of 1967, after a period when members (chiefly university and high school biology teachers) had been actively involved in attempts to abate pesticide and water pollution in the state, the Association's newsletter carried an editorial protesting the "muzzling" of biologists at a state-supported institution. "While we suppose it was inevitable," the editorial went on, "it is nonetheless regrettable that some members have apparently been deterred by their employers from taking a stand as private citizens on matters of concern to them and the Association."

Many scientists still are concerned that the flow of industry money into the universities has created a new "Lysenkoism" among the specialists working on pest controls. The national ideology, which justifies an ever-increasing production and consumption, calls for a compatible basis in science. The use of more and deadlier chemical pesticides is supported by scientists working on industry-sponsored projects; instead of lending their voices to the call for population control measures, they imply that chemicals can solve the whole problem by providing the necessary food for the teeming millions.

Rachel Carson described this dubious research, and the resultant flow of information, to a gathering of the Women's National Press Club in 1962: "We see scientific societies acknowledging as 'sustaining associates' a dozen or more giants of a related industry. When the scientific organization speaks, whose voice do we hear — that of science? Or of the sustaining industry? It might be a less serious situation if this voice were always clearly identified, but the public assumes it is hearing the voice of science."

One of the organizations to which she was referring was the Entomological Society of America (ESA). This society was founded in 1906 by biologists whose chief field of study happened to be insects. In the early 1950's, following a bitter struggle within the society, it consolidated with the American Association of Economic Entomologists. The bitterness, however, did not subside. Some members of the original ESA, particularly the systematists and morphologists, resigned in protest, while others stayed on and agitated for separation. Though the new group included many sound biologists too, the purists felt that it was overweighted with chemists, toxicologists and others whose mission was simply to destroy insects ("These people *loathe* insects," was one of the complaints. "Their life is a crusade against them!")

By the time of *Silent Spring*'s publication, the society seemed to be dominated by the economic entomologists. Among its "sustaining associates" it listed American Cyanamid Company, Gulf Oil Corporation, Monsanto Chemical Company, Shell Chemical Company, and Velsicol Chemical Corporation. The society's attack on Rachel Carson closely paralleled that of its "sustaining associates."

"The ESA's part in the controversy was most unedifying and I am certain that many members wish that the officers had at that time shown more discretion," a scientist close to the controversy has said. "A few reputations were jeopardized by individuals who took positions that later became completely untenable. A former president of the ESA was most disturbed over the position that was taken and advised against it. Also, many members who had seen the growth of the commercial interests within the ESA become dominant were quietly trying to get the ESA back on a more scientific course, even before 1962, and they had a pretty hard time."

In 1962 ESA's Committee for Public Information sent a letter to prominent scientists, asking questions about several statements made by Rachel Carson (but not indicating where the statements were taken from). "Both questions are highly loaded," one of the scientists receiving the letter has said, "and any answer which one may make to these questions can be misused by the inquiring party." Even many economic entomologists refused to cooperate with the society.

Three years later, in response to an article written by ecologist Frank E. Egler in *BioScience*, the society's leaders once more assumed a highly defensive stance. Egler was critical of that segment of the academic community which took a simplistic view of the pesticide problem: "Beetles kill trees and destroy timber; ergo, kill beetles (and forget the rest of the web)." The Entomological Society of America at its annual meeting passed a resolution condemning the "lax editorial policy" of *BioScience*, and in effect demanded the right to censor its articles. Again an uproar followed, with some members dissociating themselves from the resolution.

This sort of criticism of their own disciplines is growing as popular among the entomologists as it is among the other natural scientists. In speaking of the pesticide dispute, one entomologist has said:

> Strong criticism can justifiably be directed towards entomologists for not anticipating some of the problems which arose. For instance, resistance in insects to chemicals was not unknown; several examples were already existent. Additionally, little if any work was done on the physiological aspects of these materials in terms of the pest insects as well as the non-target organisms

in the ecosystem. Some light can be thrown on this if we consider that from the beginning, entomology in the United States has been mainly oriented to the practical needs of agriculture. The philosophy of the land grant institutions dictated these to be matters needing attention. Little if any personnel and funding were available for the more basic pursuits which would have provided the answers needed during this period of unprecedented usage of chemicals for insect control.

Another prominent entomologist spoke recently of his colleagues' attitudes since the publication of *Silent Spring*. "The book has made entomologists much more alert to the dangers of the products they use and recommend," he says. "It has initiated an all too small shift in research emphasis to biological control techniques. It has made entomologists very touchy in the face of criticism. It has placed entomologists in a position of having to defend some of the things they do."

In an earlier chapter we discussed those scientists both here and in England whose role it is to mop up behind the headlong rush of the economic scientists. Today one breakthrough in technology leads to another; what else it leads to, no one is able to say. Unfortunately, few prefer to fight this rearguard action. The results of their work appear only in the technical journals, and it is left to a Rachel Carson in the United States, a Kenneth Mellanby in England, and a C. J. Briejèr in the Netherlands to alert the public to technology's inevitable side effects.

"Let me illustrate this communications problem," says William H. Drury, Jr., of the Hatheway School of Conservation Education. "At Harvard there is a group of scientists studying baboons. These scientists are really studying people, of course, but they pretend they aren't. They've learned that when a baboon which is low in the peck order discovers a new food, none of the others will pay attention to it. But if a baboon *high* in the peck order discovers a new food, all the other members of the troop want to try it too. What we need now are people high in the scientific peck order to take up the cry aggressively against the present pesticide policies. Until then, the rest of the scientific troop won't pay much attention."

*

It is no wonder then that the chemical industry was shaken by the message of *Silent Spring*. Its able staffs of chemists and technicians, its able public relations teams and indeed its highest executives had received only one side of the story from the academics on whom they depended. The facts of insect resistance and wildlife damage came as painful surprises. Industry does not *want* controversy, of course. Many firms seek, vigorously or otherwise, alternatives to the products for which they are criticized. Nevertheless, industry in its frustration reacts rather violently to criticism. Then we are privy to the sordid spectacle of General Motors hiring detectives to pry into Ralph Nader's private life, or the chemical industry paying public relations people to link Rachel Carson with "mystics and food faddists."

Business goes on as usual. What Henry James called "the huge American rattle of gold" drowns out the warnings sounded by the ecological conscience. Five hundred salesmen and $25 million in promotion and sales expenses are concentrated each year in California's fertile San Joaquin Valley to peddle pesticides to farmers. Because of the shortage of agricultural experts, a large proportion of the farmers there rely on salesmen from the chemical companies to examine their fields and recommend the types and amounts of pesticide treatment; the salesmen rarely err on the conservative side. In 1967 the Shell Chemical Company vigorously pushed a new pesticide, Azodrin, on California's cotton growers. As a result, over one million acres throughout the state were sprayed with this new organophosphate insecticide. The Company repeated its mammoth sales promotion campaign in 1968, despite a warning from entomologists serving on the University of California "cotton recommendation committee" that Azodrin devastated populations of helpful predator species as well as the pest insects. The committee carried this warning in its 1968 bulletin: "The organophosphorous materials (Azodrin and methyl parathion) have had severe impact on natural enemies of the bollworm and their use may lead to aggravated problems."

But the gold continues to rattle. Over 80 million "bugbombs," or aerosols, are sold annually to the American public. The total U.S. sales of synthetic organic chemical pesticides (including DDT and its allies) exceeded 840 million pounds in 1966; various inorganic pesticides (arsenicals, sulfur, coppers, etc.), solvents, and spray oils added another 450 million pounds to the industry's production.

Meanwhile, the industry's promotion departments ceaselessly grind out a message which implies that man stands apart from nature, that he can successfully manipulate the environment if he takes care to follow the directions on the label. A close look at the label, however, might lead a thoughtful user to conclude that industry is not quite convinced by its own expensive message. Many pesticide labels contain the following statement: "Buyer assumes all risk of use or handling, *whether used in accordance with directions or not.*" And a U.S. Fish and Wildlife scientist is moved to insist that most of the chemical companies do not invest heavily enough in really finding out what those risks are. ("They just nail down the patent and feed it into the registration process.")

There are significant exceptions to the monolithic self-righteousness of industry's response to *Silent Spring*. Segments of the industry today are actively seeking to improve their public image while searching for pesticides not quite so biocidal. A recent count disclosed that only seven of the country's 50-odd leading chemical manufacturers continue to produce DDT. To meet USDA's more stringent registration demands, many of the companies have built new research facilities, one of which has been called by the *Wall Street Journal* "a million dollar monument to Rachel Carson."

One of the most prominent examples of an enlightened approach by industry may be found at Midland, Michigan, where the Dow Chemical Company woos conservationists while fending off critics of the Vietnam War. Dow not only makes extensive investigations into the side effects of its new products, but it places advertisements to spread the news in conservation publications. One such full-page advertisement described the development of a new herbicide called Tordon; Dow carried out life-cycle studies on the food supply of the kingfisher to make certain that Tordon, seeping into waterways, would not affect the aquatic food chains which ultimately support this fish-eating bird.

Another recent Dow development is Plictran, a miticide so specific in its action that the company's claims go beyond the fact that it is harmless to fish and wildlife; Dow reports that its miticide destroys only plant-eating mites, while sparing the predaceous mites that help to control harmful insects around apple trees! In such fruitful research, rather than in personal attacks on concerned scientists and conservationists, lies the chemical industry's best hope of convincing the public of its good intentions.

14. The Message

WHAT MESSAGE finally gets through to the people who must make the decisions about our pesticide policies? Alfred L. Hawkes of the Audubon Society of Rhode Island tells a revealing little story.

"Our concern over the effects of pesticides on the environment has made practically no impression on any of the officials or agencies in a position to take corrective action within the state," Hawkes says. "So recently I offered a copy of Robert Rudd's book, *Pesticides and the Living Landscape*, to one of the most important of these officials. He countered a week later by handing me a copy of Jamie Whitten's book, *That We May Live*."

Rudd's book has been described by one ecologist as "a sequel to, and a vindication of, *Silent Spring*." It occupies, as we have seen, a prominent place among the documents that have helped to set the record straight in the pesticide controversy. *That We May Live* plays another role in the controversy, and a close look at the background of this curious book sheds light on a whole area of American politics.

Jamie L. Whitten, its author, is a congressman from Mississippi and the chairman of the House Appropriations Subcommittee on Agriculture. In 1964 Whitten requested the Surveys and Investigations Staff of the Appropriations Committee to "conduct an inquiry into the effects, uses, control, and research of agricultural pesticides, as well as an inquiry into the accuracy of the more publicized books and articles which increase public concern over the effects of agricultural pesticides on public health."

Whitten, a spokesman for agricultural interests in Congress, was understandably alarmed by the rising tide of public opinion against the current pesticide policies. The impact of *Silent Spring* had been followed closely by the recommendations made by President Kennedy's Science Advisory Committee. At the time that Whitten ordered a study, Senator Abraham Ribicoff's Subcommittee of the Senate Committee on Government Operations was gathering voluminous testimony on pesticide use, much of it highly critical of present policies. (The so-called Ribicoff Report, emphasizing the lack of vital knowledge about the effect of pesticides, was released in 1966.)

Moreover, USDA officials were restive. For one thing, they were anxious to counter the bitter criticism aimed at their past pesticide blunders, particularly their massive "eradication" programs. For another, these officials were concerned about a decline in the demand for their services in agricultural areas. The solution here, in their minds, was obvious; it was to broaden USDA's base by assuming the Department of the Interior's role as the conservation arm of the Federal Government. Whitten frequently has tried to support this power grab by chopping away at the appropriations granted Interior for research on pesticide-wildlife problems.

The committee staff interviewed over 200 scientists and physicians, as well as state and government officials, conservationists, and industry people. In 1965 it issued its report (which was immediately reprinted by the National Agricultural Chemicals Association). The report made a certain impact on the public, since some newspapers gave prominent space to the news that a "Congressional Committee" had criticized Rachel Carson. Fortunately, the news stories generally quoted from that section of the report containing (as we shall see) a statement which exposed the entire project for what it was: an apology for the status quo in pesticide use.

Much of the early part of the report was devoted to a critical review of *Silent Spring*, prepared by the staff members and based on interviews with "scientists and physicians" whom the report did not identify. (Some of the scientists interviewed later refused to allow their names to appear when Congressman Whitten used the report as a basis for his book; others, working for the government, felt they were not in a position to turn down a congressman's "request" for the use of their names. In any case, Whitten carefully avoided saying that those who were interviewed agreed with the book's conclusions — but the reading public was left with the impression that they did.)

The report's review of *Silent Spring* read, in part:

> The staff was advised, by scientists and physicians, that the book is superficially scientific in that it marshals a number of accepted scientific facts. However, it is unscientific in (a) drawing incorrect conclusions from unrelated facts, and (b) making implications that are based on possibilities as yet unproved to be actual facts . . .
>
> For example, the unsupported statement was made that the allegedly serious pollution of air, earth and water with chemical pesticides initiates a "chain of evil" which "is for the most part irreversible." Persistent pesticides remain in the soil for long periods, but there is no evidence that they cause a "chain of evil" . . . The author then made the statement that chemical pesticides may "pass mysteriously by underground streams until they emerge and . . . combine into new forms that kill vegetation, sicken cattle and work unknown harm on those who drink from once-pure wells." Scientists advise that any proof of pesticides combining into new forms is lacking.*

* Having written that there are no proofs of pesticides combining into new forms, the staff members contradict themselves elsewhere in the report (page 194): "Some peach pickers in California were made ill, in August, 1963, by touching the leaves while picking the fruit; the trees had been correctly sprayed with parathion, which under the usual climatic conditions prevailing, had changed to a more toxic compound."

An analysis of the Whitten Report reveals that it so closely follows the chemical industry's defensive promotional line that it makes an excellent primer on the subject. All the discredited old chestnuts are here:

- The report lays great stress on the fact that pesticides are applied to only five per cent of the nation's land. There is no mention made of the fact that pesticide residues have been detected *everywhere.*
- The report recognizes only that pesticide hazards result from "improper use." Even before the report was written, studies on the recycling and magnification of toxic residues through food chains made nonsense of this limitation.
- The report claims that the real controversy is not over whether the environment has been contaminated by pesticides, but whether they should be used at all; the implied alternative to pesticides would be a failure to feed our people and meet our foreign commitments. This is the simplistic message spread by the agri-chemical industry to convince the public that the philosophy of *Silent Spring* leads only to food shortages and higher prices.
- The report implies that pesticides in many cases have caused an increase, rather than a decrease, in wildlife populations. This sort of reasoning causes wildlife management experts to tear their hair. Where wildlife populations have increased because of pesticide use, it generally is because pesticides simplify the environment so that certain populations increase to pest proportions. Thus blackbirds and starlings increase about farms, while more beneficial, insect-eating birds have disappeared; and deer, rabbits, and small rodents increase to unwieldy numbers where their predators have been poisoned.
- Elsewhere the report admits some damage to wildlife, but insists that human health is not jeopardized. This part of the report gives support to the philosophy that known toxic substances should not be restricted in the environment until their

harmful effects have been proven scientifically. (The difficulty of establishing this evidence for human beings is ignored.) No support is given for the contention that known toxic substances should not be put on the market until their safety can be established.

• The importance of new pesticides is equated with that of new drugs. But the report fails to mention that a great many new drugs have been found harmful to patients on whom they are used, and many have been withdrawn from use after the tragedy has occurred. Yet drugs are harmful only to the individual on whom they have been used; the widespread application of pesticides makes guinea pigs of us all.

• The report lays great stress on the minute traces of toxic chemicals found in the human body. It makes no reference, however, to the fact that toxicological studies of many of the new compounds are either nonexistent or in a very early stage of development.

• The report makes this judgment: "The individuals who oppose the use of pesticides claim that pesticides are harmful; whereas, the individuals who support the use of pesticides, claim that pesticides are harmless. Because no one at the present time can prove who is correct, any discussion of long-term effects on minute traces of pesticides on human beings would be fruitless."

It was this extraordinary statement which was quoted in the press, revealing the quality of the report even to the casual reader.

Whitten, convinced that pesticides had become "an absolute necessity to our way of life," resolved to pass on his feelings to the American public. This he did by turning the stuff of the "Whitten Report" into a book of his own. Published in 1966 under the title *That We May Live*, it is rather folksy and engaging in parts, with the man behind the title showing through rather more clearly than in most books "written" by politicians.

But Whitten's book reflects the uncritical assimilation of material which marred the original report. It is clearly an answer to Rachel Carson, and she is well thumped during long passages in which the author occasionally pauses to pay tribute to "her almost unmatchable prose and the resonance this arouses in one's own spirit."

Printed matter so unscientific, whether it appears as a book or a government report, contributes nothing of value to the public debate. If scientists are making vital discoveries about pesticides and their action in the ecosystem, the facts ought to be put on the record. Only then will society be able to establish a pesticide policy based not on pest control techniques but on fundamental biological principles.

Part IV

The Uncertain Defenders

15. Washington Begins To Stir

"THE CORPORATION's convenience has been allowed to rule national policy," said the 1965 report of the President's Science Advisory Committee, *Restoring the Quality of Our Environment*.

This important document got at the crux of the problem. Firmer legislation may reduce industry profits. The switch from persistent to more acceptable pesticides at first may cause the farmer added labor and expense. Private convenience clashes with public welfare. Who can stop the farmer from applying ten times the recommended dosage of pesticides to his fields?

By 1965 many bureaucrats in Washington (scientists and nonscientists alike) were trying to come to grips with the pesticide problem. Obviously, they had been made uneasy by the message of *Silent Spring* and its repercussions in public life. There was a general awareness that something had gone wrong. But it is a measure of the often myopic, always ponderous, nature of the Federal Government that, although highly placed men like President Kennedy and Secretary of the Interior Udall spoke out early against pesticide abuses, the general response in Washington remains ambiguous.

• Regulations pertaining to pesticide use on lands controlled by the Federal Government have been tightened up in many cases. Yet only five per cent of the pesticides used in the United States are applied by federal departments. The rest of our massive production for domestic markets is used at the discretion of the state and local governments or of private citizens. (U.S. sales of pesticides exceed 1.25 billion pounds a year, of which 400 million pounds are sent abroad.)

• The Federal Government's pesticide policy remains fragmented because each cabinet-level department owes allegiance to separate industries and causes. The Department of Agriculture exists to help (and listen to) the farmers; the Department of the Interior serves many masters, including stockmen, miners, and the vacationing public; the Department of Health, Education and Welfare, through the Public Health Service, wages war against disease-carrying insects. Inevitably, their pest control policies sometimes conflict.

• The fragmentation is further magnified within the departments themselves, where subsidiary agencies fall into family squabbles. Interior's predator control programs, for instance, may damage a wildlife propagation program carried out elsewhere in the department.

• The Federal Government exercises no control over the chemical industry's research in this vital area. Laws banning the more hazardous pesticides such as DDT would force industry to make greater efforts to develop alternative materials. Yet even a bill providing for the inspection of pesticide manufacturing plants (to help prevent such environmental disasters as the Mississippi fish kill) has been snuffed out repeatedly in Congress; the industry expressed its "complete and vigorous opposition" to this sort of legislation, and had the muscle to work its will on congressmen. The feeling in many high places is that the chemical industry, left to its own devices, eventually will police itself. "The economics of the country dictate this," one USDA official has said.

Into the regulatory void, affectionately referred to by industry people as "laissez faire," flows that deluge of propaganda which so often numbs common sense. ("When weeds win, people lose," the ads intone.) The public, either directly or through taxes to its local, state, and federal governments, pays well over $1 billion a year for pesticides. Industry, therefore, fights government regulation with great zest; $1 billion generates a flood of self-righteousness, and the cry for reform unfailingly evokes the Bolshevik Revolution in the minds of some industry people.

> The 1965 PSAC report, an exhaustive study which never received the attention to which it was entitled, concluded that environmental pollution by pesticides could be materially reduced, in certain cases by a half, with no loss in efficiency in pest control by making use of methods already available. To do so will require recognition that neither 100% control of pests on a crop nor eradication of a pest generally is required to prevent economic loss. The public at large, farmers and extension and research workers will need to be convinced that virtually complete reliance upon chemical pesticides is a mistake, and that bioenvironmental controls offer great promise. To date this promise has not been realized. Our recommendations are directed to the wider use of those bioenvironmental methods already available, to the discovery of additional methods, and to assuring an adequate supply of trained manpower to accomplish these needs.

The Federal Government, however, has very little jurisdiction over industry research, and none over the application of pesticides once they have been bought by the private citizen. At present, aside from the registration of chemical compounds, it is best equipped to regulate pesticide application by advice and example. We have seen that sometimes, in the past, neither has been salutory. As for federal advice, the best intentions are generally present on the highest level of government, and often it is simply a matter of knowledge and intentions dribbling down to the workers who are in touch with the public. And as for federal example, a great deal of mischief has been done by the failure of various government agencies to agree either on a general pesticide policy, or on an approach toward a specific problem.

The problem of inter-agency cooperation was one that concerned President John F. Kennedy when he took office in 1961. This concern manifested itself in the pesticide field with the creation of the Federal Pest Control Review Board on October 1 of that year. This was an advisory board, composed of two members from each of the federal departments engaged in large-scale pesticide programs — Agriculture; Defense; Health, Education and Welfare; and Interior. The idea was to provide a systematic joint review of all major federal pest control programs, coordinating the activities of the four departments. But this board was not very effective. The 1963 Report of the President's Science Advisory Committee recognized this deficiency in the following recommendation: "Provide clear assignments of responsibility for control of pesticide use."

Accordingly, by a charter signed by the secretaries of the four departments concerned on July 27, 1964, a new committee was created. It is called the Federal Committee on Pest Control (FCPC). As before, each department assigns two members to represent it on the committee. To modify the government's habit of acting, as someone has said, "retroactively rather than prospectively," all federal pest control programs are scrutinized in advance by the committee's members. For instance, if the Forest Service is planning to spray an area for the control of spruce budworm, the two USDA members will go into detail for their fellow committee members about the proposed plans of the application. They will discuss dates, dosages, and areas, as well as safety precautions. If the Fish and Wildlife Service has a wildlife program in that area, the two members of the Department of the Interior will make that known, and the program then will be modified or abandoned. Although the committee has no mandatory power, all of its recommendations have so far been accepted by the agencies concerned.

Ideally, the FCPC is a sort of federal watchdog, creating a dialogue on pesticide problems among the departments and setting up a system of checks and balances.

"There is a whole new attitude about pesticides in the Federal Government," an official of the Department of the Interior has said. "The various departments are moving as fast as they can now, with what's available to them, toward more acceptable alternatives. USDA, for instance, has become very cooperative. Now they'll throw their hat in the door and come in after it with an apology: 'Gee, fellas, we've tried everything else — can we *please* use this?' "

One critic of the FCPC has pointed out the handicap under which its members work. "They have the awkward function of serving as advocates of their own agency's projects as well as acting as judges of the plans of other committee members' agencies," wrote Shirley Briggs in the *Atlantic Naturalist*. "This violates a basic principle of good government. Such a committee should be recognized as a quasi-judicial body, and given protection from conflicting functions. A higher authority should be set up to pass final judgment. Regardless of the qualifications of individual committee members, we have given them an impossible task. So, when a man from the Public Health Service insists that his agency must use a persistent residual chemical like DDT, his technical reasons based on epidemiology or the habits of a particular disease vector may be accepted without question by other committee members, who are not in a position to call for additional disinterested evidence."

The committee's staff does not minimize the difficulties in pulling together the disparate arms and aims of an enormous government. "It's even hard to get them to agree on a statement sometimes," one of them says. "USDA may want to say, 'This pesticide poses no hazards,' and the people at HEW may want to add, 'when properly used.' "

To make its task of overseeing the government's pest control activities more manageable, the committee has established a number of subcommittees, dealing with pesticide monitoring, research, safety in marketing and disposal, and public information.

The monitoring program is of special interest. USDA, HEW, and Interior cooperate in providing information which eventually may let us know to what extent man and his environment have become contaminated. The Food and Drug Administration carries out its market basket studies, USDA analyzes soils, Interior watches fish and wildlife and water resources, and the Public Health Service focusses on human beings in its community studies.*

The Federal Committee on Pest Control also coordinates research and public information. In the latter field it has, for good or ill, worked to unify the statements made about the pesticide problem by various departments within the Federal Government. The underlying danger, of course, is that the committee may tend to smooth over the differences in viewpoint instead of trying to solve them. The public information subcommittee provides speakers' kits for government employees assigned to make talks on the subject of pesticides.

The FCPC has become concerned about the waste disposal problem. "We've discovered that no one was really responsible for pesticide waste disposal — dumping offshore and things like that," a staff member has said. "We're trying to bring people together on this. One of the first groups we went to was the National Agricultural Chemicals Association several years ago to get them to tell us what industry is doing about its disposal problem — especially their containers. Industry is alerted on this now."

If the pesticide problem was created to a great extent by a breakdown in communications, the Federal Committee on Pest Control may still provide a valuable service by building up a communication system within the Federal Government. It would be wise to have one hand know what the other is doing; too often, for instance, a federal wildlife program with a heavy investment in both time and money has been wiped out because of some other department's concern with a particular pest. In

* The monitoring program will be discussed more fully in Chapter 18.

the absence of firm restrictive legislation on pesticide use, the Federal Government must provide sound advice and constructive leadership for both industry and the states. The need, as we shall see, is critical.

16. Predators and Poisons

DURING THE LAST MONTHS of her life, Rachel Carson's attention often was fixed on a special aspect of pest control. This aspect was symbolized for her by the mindless, scattergun techniques of slaughter employed by the Federal Government's Predator and Rodent Control program. Much of her correspondence during those months stimulated or reflected this concern. There was, for example, a letter from a little girl in Boulder, Colorado, telling of the prairie dogs she used to see while horseback riding with her parents. Having noticed one day that they were gone from their usual haunts, she asked about them and was told that "the exterminators" got them.

"I don't think they should be killed," the little girl wrote to Rachel Carson. "Do you?"

Of the reasons for that particular slaughter Rachel Carson might have been ignorant. But she had become convinced that, in most cases of predator and rodent control, she would have had to reply to the little girl's question with an emphatic "NO!" She based her reservations about the program both on reports from distinguished biologists and conservationists among her acquaintance, and on her own wide reading. A book that impressed itself on her mind was Farley Mowat's *Never Cry Wolf*, which recounts the savage campaign of "extermination" carried on against the wolf by the Canadian government. Only four months before her death, Rachel Carson wrote the following words to a British conservationist:

"On reading Farley Mowat's book I myself had felt this was a shocking revelation of the archaic philosophies that direct the handling of this matter by the Canadian Wildlife Service. The so-called predator control activities of our own Fish and Wildlife Service are no better."

Hers would have been a strong voice in any campaign to bring the government to its senses. Nevertheless, at the time of her death, the current of opinion was beginning to turn against these "archaic philosophies" that were hangovers from a more primitive time when predators were branded "murderers," "criminals," and "vermin." (Robert L. Rudd referred to this viewpoint among predator control officials as "the Mother Goose Syndrome.")

Traditionally, of course, certain animals have received "protection" — in most cases because men happened to think that they tasted good. Other animals gradually earned this preferred treatment only by being pushed to the edge of extinction. But in the past century the federal and state governments have been under unrelenting pressure from farmers and stockmen to eradicate whole animal populations that were thought to cause them financial losses. The result of this pressure was the creation of a "science" of predator control that, through a sort of inertia, escaped finally from the control of those responsible for it.

It is sometimes overlooked, even by conservationists, that predator control is a part of the pesticide problem. A predator, in the eyes of farmers and stockmen, often is a pest *per se*. The poisons with which control experts kill predators (as pests) are classed as economic poisons, or pesticides. And in predator control we find most of the abuses for which other pest control programs have been criticized in the past.

Referring to predator control in 1967, G. G. Simpson, Professor of Vertebrate Paleontology at Harvard, wrote that "Under this euphemistic label our tax supported wild life agencies are largely devoted to the extermination of wild life. (In their jargon, 'control' means complete annihilation, just as the good ladies of Evanston, Illinois, mean absolute prohibition when they say 'temperance.')"

This, of course, is the fallacy of modern predator control methods. They are wasteful both of wildlife and public funds. Predators *do* cause some destruction, and occasionally some control is justified. More often, however, the destruction is exaggerated. Fishermen who overfish their waters and stockmen whose cattle and sheep have overgrazed their lands prefer to attribute their dwindling profits to "predators" rather than to their own stupidity.

Even where excessive predation has been documented and predator control is justified, the authorities continue to abuse their license. All-out war against far-ranging animal populations can never be justified. The bounty system and extensive government hunting programs, although proved to be biological and financial blunders, at least may be defended on the grounds that they are confined in theory to the target species. (The repeated frauds of bounty hunters, who substitute the carcass parts of easier prey than the designated animals when applying for their bounties, tend to weaken even this defense.) But we are concerned with poisons here. The control of predators by poison ranks as one of the sorrier chapters in the history of man's relations with what Henry Beston called those "other nations, caught with ourselves in the net of life and time."

Is the Federal Government at fault? No, and yes. The subject of predator control makes an enlightening introduction to any study of federal pesticide policy because it explains in part the difficulty, even under the best of administrations, of altering discredited practices. In general, the Department of the Interior, and in particular its Fish and Wildlife Service, are favorably looked upon by conservationists. But the department harbors within it a variety of agencies which sometimes work at cross-purposes to one another; thus the dam-building activities and drainage of the Bureau of Reclamation, and the mining practices approved by the Bureau of Mines, may undo the work of the department's wildlife specialists.

Furthermore, the department inherited certain problems in 1939 that it has never completely solved. At that time the Biological Survey was removed from the Department of Agriculture, and the Bureau of Fisheries from the Department of Commerce, and the two were combined in the Fish and Wildlife Service under the Secretary of the Interior. Earlier in the century the federal predator control programs had been surpervised by the Biological Survey.

The considerable advances that had been made by the U.S. Government in conservation education and practice during the 1930's were unavoidably curtailed by the outbreak of World War II. They received an even greater, if less explicable, setback in 1952 with the appointment by President Dwight D. Eisenhower of Douglas McKay as his Secretary of the Interior. McKay's comprehension of conservation was best reflected by his characterization of conservationists as "punks"; his program consisted chiefly of leasing away public lands and driving out of the department such dedicated wildlife specialists as Alfred M. Day, Clarence Cottam, and Durward L. Allen.

Meanwhile, the predator control people found that their lives had been made a great deal easier by the deadly poisons that arrived on the market after World War II. Their chief weapon became sodium fluoroacetate, which is commonly called by its experimental number, Compound 1080. This rodenticide, derived from an organic salt, is a stable, water-soluble compound that is tasteless and so cannot be detected by the animal consuming the bait. Though 1080's action is delayed sufficiently so that the animal does not sicken before it has eaten the prescribed lethal dose, this poison is generally considered to be the most potent now used widely in pest control. No wonder that 1080 was hailed as enthusiastically as DDT!

Yet, despite the new poisons, predator control methods grew less specific. Whereas, for instance, poisoned baits had once been placed by hand, now they were broadcast indiscriminately over the countryside from planes. Poisoned bait, dropped from the air to control gray wolves in isolated regions of Alaska, killed bears and valuable fur-bearing animals. Poisoned grain, carelessly distributed during a campaign against meadow mice, killed thousands of geese in the Tule Lake area of Oregon and California in the late 1950's.

Moreover, incidents of "secondary poisoning" became better publicized. This phenomenon — the poisoning of an animal which has eaten the carcass of another animal already poisoned — grew widespread as the tasteless, delayed-action properties of such compounds as 1080 permitted the initial victim to consume large quantities of the poison. After it died, there remained lethal dosages in the viscera and other parts usually preferred by scavengers. Eagles fell victim to secondary poisoning following control programs directed at the prairie dog; coyotes died after eating poisoned mice and other rodents. The target animal, dying, remains a reservoir of death.

It is very much to the credit of Stewart L. Udall, perhaps the finest Secretary of the Interior the nation has had, that he initiated a review of all of his department's pesticide programs — including predator control. In 1964 his resources program staff laid down a criterion for the evaluation of a specific pesticide that many other federal and state agencies have yet to approach: "As a general principle, the case against a given product does not require evidence equal to that in favor of it. All that is required is sufficient evidence to establish reasonable doubt as to safety."

Interior's concern with the pesticide issue did not begin with the publication of *Silent Spring*. For a number of years the department had been accumulating information of a disturbing nature about the effect of pesticides on the environment. When Rachel Carson suddenly brought the issue into the open, Udall had a fund of solid information on which to base his stand; the fact that he alerted his assistants to keep up with every development relating to the publication of *Silent Spring* and its aftermath made it easier for him to establish a sensible policy for his department.

In 1964, therefore, Udall approved a memorandum ordering each of his chief assistants to assume responsibility for all pesticide uses in programs under their supervision. He reminded them that they must seek out the advice of technical experts, either in or out of government, to assure themselves that the assessment of environmental effects was complete.

"The standards of this Department in the use of pesticides should be higher than any others," Udall wrote in his memorandum. "We should set an example which others can follow."

Udall's memorandum was not bureaucratic hogwash. Interior has consistently tried to find answers to difficult questions in the pesticide field (see Chapter 9). Udall banned the use of DDT on Interior lands when alternatives will serve the purpose. Even so, incidents of spraying within the national parks have aroused criticism on occasion. A telegram to Udall in 1963 from Richard M. Leonard, then secretary of the Sierra Club, protesting the spraying with malathion for needle miners in Yosemite National Park, sums up the reasons for the conservationists' concern.

"Strongly support your halt of poison in national parks," Leonard wired. "Needle miners have existed a million years under control of predators and parasites. Some natural areas must be left on earth to permit continuation of original evolution. Unless parks maintain strict natural preserve conditions, then basic research on primeval ecology which can be so valuable to commercial forest ecologists elsewhere becomes impossible . . . The decision must be on basic park ecology, not limiting control temporarily by acreage. No aerial spraying should be permitted on national parks."

The spraying still took place. It was justified by Interior officials not on the basis of potential serious damage to the forest, but because spotty foliage might offend the sensibilities of tourists. There is an increasing reaction by conservationists today against this sort of cosmetic treatment of forests with powerful poisons.

Udall also had inherited a predator control program mired in mindless inertia. In 1962 he initiated a review of his department's predator and rodent control operations. Certainly his inquiry was not dictated by political considerations. Predators and rodents have few influential friends. The health of human beings is not jeopardized by such programs (though often their animal pets come to grief). Farmers and stockmen, as well as the legislators who guard their interests with such zeal, exert considerable pressure for more extended control programs.

But in appointing a committee of distinguished biologists and conservationists, headed by A. Starker Leopold,* to review these programs, Udall was responding to a concern that has little to do either with human health or economic interests. The fisherman with a hankering for a couple of fried trout, the housewife plagued by mice or ants, even the hunter with a trophy room to fill — each has a reason for taking life. But there is a growing revulsion against unnecessary and indiscriminate slaughter.

The attention of the Leopold Committee, as it came to be called, was directed to the branch of Predator and Rodent Control of the U.S. Fish and Wildlife Service. PARC provided an instructive example of a government agency that had slid out from under the thumb of its parent department. Stationed in the far West, where they found the widest opportunity to pursue their grim calling, most of PARC's 800 employees remained in closer touch with farmers and stockmen than with their superiors in Washington. From all accounts these professional "gopher-chokers" were not above drumming up business. They poured, year after year, an unvarying message into the grow-

* The other four members of the Secretary's Advisory Board on Wildlife Management were Stanley A. Cain, Clarence M. Cottam, Ira N. Gabrielson, and Thomas L. Kimball.

ers' ears: wild animals destroyed their crops and livestock, control was a necessity. Since the government paid for the programs, farmers and stockmen went along with them. And PARC consolidated its little empire.

The Leopold Committee came to some emphatic conclusions. "Control is far in excess of the amount justified," its report said. "Many animals which have never offended private property owners or public resource values are being killed unnecessarily."

Control had become an end in itself. Most control programs were carried out without economic justification. Control was based almost exclusively on subjective judgment, and not on data supporting the extent of damage inflicted by wildlife. For instance, cattlemen supported the indiscriminate poisoning of entire prairie dog towns, although these rodents never have been proved to be a primary cause of range deterioration; in fact, rangeland studies indicate that rodents and rabbits prosper chiefly on range that already has been overgrazed by cattle. Sheepmen, of course, are notorious for blaming their losses entirely on predators. (The coyote, detected in the act of eating a lamb which may have been born dead or a sheep that may have died of natural causes, is invariably branded a "killer.")

"The federal program," said the Leopold report, "is to a considerable degree designed by those who feel they are suffering wildlife damage. Too often PARC personnel support control decisions without critical appraisal. At times they even solicit requests for control and propagandize against predators. There is no mechanism to assure that the positive social values of wildlife are given any weight or that control will be limited to minimal needs."

One of the most objectionable features of the federal control program was that sheepmen, already subsidized by the permission to graze their sheep on federal lands, were granted further subsidies by the practice of tax-supported control programs on that land. During the years 1941-1962, the cost of control in national forests exceeded the loss of sheep from *all* causes. In eighteen national forests in California, records for 1962 showed that $90,195 was spent for predator control, while sheep valued at only $3,501 were lost.

"Admittedly, losses would have been higher without coyote control," the Leopold report said, "but how much of this control is really justified? Most of the 64,743 sheep now grazed on national forests in California are concentrated in the north and east, yet traditional coyote control programs are continuing in other areas where few if any sheep are now pastured and where recreation is the primary use."

The report was especially critical of the poisoned bait program directed at rodents. This, in part, was what it had to say:

> In 1963, PARC distributed approximately 250,000 pounds of treated bait for rodents; 150,000 pounds of this was treated with 1080. It is curious that PARC will distribute great quantities of 1080-treated grain (sometimes by airplane, as in forest reseeding projects) in the *same* areas where it takes elaborate precautions to protect carnivores other than the target species. Secondary poisoning can have heavy impact on small carnivores and birds.
>
> Even where there are no sheep and where coyote damage is negligible, the coyote has been extirpated as a secondary result of rodent control programs. In addition to the coyotes, badgers, bears, foxes, racoons, skunks, opossums, eagles, hawks, owls and vultures are exposed to possible secondary poisoning. In some localities 1080 is used to kill rodents that may hinder forest production with resultant exposure of many animals to the poison.

In submitting its report to Secretary Udall in 1964, the Leopold Committee urged a drastic reduction in the scope of the control programs, more emphasis on local control and the elimination of destructive "individuals," and further research to develop more specific methods of controlling pests in areas where programs have been proved to be necessary. Acceptable ways had been found to control game species when they became pests; why not nongame species?

"The federal government should be setting an example," the committee said. "Instead, the Branch of Predator and Rodent Control has developed into a semi-autonomous bureaucracy whose function bears scant relationship to need and scientific management."

To push this "bureaucracy" in the right direction, as well as to improve its image, the committee suggested that PARC's name be changed. Accordingly, in 1965, a reorientation of PARC began at Interior under a number of new faces (including Stanley A. Cain, a member of the Leopold Committee who took a leave of absence from his position as chairman of the Department of Conservation at the University of Michigan to become Assistant Secretary of the Interior for Fish and Wildlife and Parks).* PARC, discredited, withdrew for its face-lifting and reemerged as the Division of Wildlife Services.

Almost immediately, Cain signed an order suspending the poisoning of prairie dogs in parts of South Dakota because of the presence there of the black-footed ferret, a species known to be on the edge of extinction. This ferret preys on prairie dogs. Under Cain's order, 1080 was not to be used at prairie dog towns until observers had determined that there were no black-footed ferrets in the area. The Department of the Interior also assigned a biologist to study the ferrets' hazards and needs. The department also refused to take part in South Dakota's campaign against foxes because the state had failed to prove that foxes, rather than food supplies, had caused a decline in the pheasant population.

Many conservationists remain unsatisfied. E. Raymond Hall, a Kansas biologist, has pointed out that policy changes in Washington often are ignored several thousand miles away in the field. "Although the attitude at the three or four highest levels of administration in Interior is enlightened and therefore offers

* John S. Gottschalk became Director of the Bureau of Sport Fisheries and Wildlife, and Jack H. Berryman became Chief of its new Division of Wildlife Services.

opportunity for needed change in policy . . . employees in the field are not about to change," Hall says. "They know that the average tenure is short for the makers of policy; many of them have served under more than one director, assistant secretary, and secretary, and aim to serve under several more."

The government's hand in poisoning campaigns against the black-tailed prairie dog stirs increasing concern. These attractive little rodents are a part of the legend of the Great West; their towns often extended for miles across the plains. But cattlemen, whose herds had devastated so much of the original range, looked upon the prairie dog as a competitor for what was left. They induced the government to initiate their vast poisoning programs, beginning with strychnine-soaked grains, at the end of the last century, and continuing through the years with more sophisticated poisons such as thallium sulphate and 1080. Millions of acres of dogtowns have been wiped out.

The old PARC workers resented the new policy directives that came down to them from Washington. John Madson, a prominent conservationist and writer on outdoor subjects, reported in 1968 that many state game officials in the West began to complain about the continuing campaign by the old PARC crews to "exterminate" the prairie dog. Asked how many acres of dogtowns remained in his state, one local official replied: "I don't know. We've never officially pinpointed our dogtowns, and I hope we never do. If such a map fell into federal hands, it would be an ideal guide in wiping out our prairie dogs."

Others told Madson that disgruntled federal workers had stirred up their rancher friends against the department's more enlightened policies, saying in effect: "We're not supposed to talk to you anymore, or offer to help you wipe out the varmints that are costing you money. Sorry, old pard, but it's orders from Washington."

Madson concludes that, if the federal workers remain overzealous in their war against the prairie dog, they show little enthusiasm for fulfilling their concomitant obligation — the protection of the black-footed ferret. The observation of dogtowns to make certain that poison is spread only in the absence of ferrets is carried out usually with an appalling lack of either concern or competence. This is especially serious in South Dakota, which seems to be the last refuge of the dwindling ferret population. Madson writes of an interagency meeting there between federal control workers and state wildlife officials:

"One federal agent stated that prairie dog poisoning came first — and that research on ferrets and prairie dogs was of secondary importance. When a state wildlifer naively commented that the research should come first because South Dakota might be the last place in the world where the black-footed ferret could be saved, there was laughter."

The dedicated men in command at the Department of the Interior obviously are convinced that a new day has dawned in vertebrate pest control. However, as Madson writes, "that conviction emanates from Washington, D.C. Out home on the range, where colored oats wear a toxic bloom, it's another story."

17. The Toxic Field of Mars

IF THE INTERIOR DEPARTMENT sometimes loses control over the application of pesticides made by its subsidiary bureaus, another federal department may pursue pesticide policies wholly outside the usual framework of governmental control. The Department of Defense (DOD) in this case, as in many others, is unique. It works on the principle that pesticides may be used in a crusade to destroy human beings, as well as to save or nourish them.

Defense, of course, deals in many poisons. Much of this activity is secret, except when disaster strikes, as it did in 1968 when something went wrong at DOD's Dugway Proving Ground in Utah. The Army operates the proving ground to conduct field tests in chemical and biological warfare. In March the Army began tests on a nerve gas, releasing it in the area from 155-mm. shells and low-flying planes. The nerve gas, or "agent," was a member of the organophosphate group. The tests were not considered hazardous, but apparently a malfunction of the spraying equipment caused one of the tanks containing the agent to remain open on the plane for some seconds after it should have been closed. Thus the poisonous agent was sprayed at an altitude higher than expected, from where winds carried it into nearby valleys. The next day, sheep grazing nearby in Skull Valley began to die. Though the army stopped its testing, more than 5,000 sheep succumbed to the deadly gas and the chief of staff of a local hospital expressed fears for the safety of sheepherders, ranchers, and Indians who inhabited the area. Government tests disclosed no immediate ill effects among local residents, and measures were put into effect to prevent further accidents.

This, however, was an accident, and accidents are a readily accepted part of man's commitment to live dangerously. Of more concern to the scientific world was DOD's decision to lay waste the countryside of our ally, South Vietnam, with herbicides. This operation was designed to expose enemy positions and indirectly save the lives of American and South Vietnamese soldiers. In closed debate, the benefit clearly outweighed the risk (which the Department of Defense considered to be negligible, anyway). More open scientific debate on the subject might have served as a reminder to the generals that the United States Government, following policies it considered to be in its own political and military interests, has made mistakes in the past. Consider the following two quotations.

President Dwight D. Eisenhower, speaking on the Atomic Energy Commission's tests of hydrogen bombs, October 24, 1956:

> The continuance of the present rate of H-bomb testing, by the most sober and responsible scientific judgment . . . does not imperil the health of humanity.

President Lyndon B. Johnson, speaking of the new Nuclear Test Ban Treaty on October 12, 1964:

> This treaty has halted the steady, menacing increase of radioactive fallout. The deadly products of atomic explosions were poisoning our soil and our food and the milk our children drank and the air we all breathe. Radioactive deposits were being formed in increasing quantity in the teeth and bones of young Americans. Radioactive poisons were beginning to threaten the safety of people throughout the world. They were a growing menace to the health of every unborn child.

What had happened, in less than eight years, to reverse the attitude of our highest office on a matter of such profound importance to mankind? As Barry Commoner has pointed out in his book, *Science and Survival*, great secrecy surrounded the early activity of the Atomic Energy Commission. The public (scientist as well as layman) had no alternative but to accept the government's assurance of "no hazard." Then, in the middle 1950's, the AEC began to release certain information to the scientific community. Independent scientists swiftly uncovered the hazards of nuclear fallout. Linus Pauling, and others, suggested the frightening consequences inherent in the release of radioactive materials to the environment. Pauling's conclusions were at first disputed by government scientists. Later, when the evils came home to roost in, among other forms, the thyroid nodules developed by children both in Utah and in the Pacific who were exposed to blast radiation, the independent scientists were vindicated. Today even government scientists seem to be aware that hydrogen blasts are hardly therapeutic.

Yet, with glib unconcern, DOD turned loose a withering blast of herbicides on the Vietnam forests and croplands in the late 1960's. Through September, 1967, DOD admitted that over two million acres of that ravaged country had been sprayed with defoliants, chiefly 2,4-D and 2,4,5-T. And the rate was just beginning to accelerate. According to a USDA booklet issued a month later, "Some domestic users of 2,4,5,-T in 1967 have had to substitute other chemicals for such purposes along railroads. However, supplies available for domestic users, including formulator and dealer inventories of 2,4,5-T, were sufficient to avoid serious domestic shortage for the 1967 season. As to 1968, fulfillment of current contracts for military use of 2,4,5-T will drastically curtail supplies for domestic use."

The Department of Defense, in effect, commandeered much of the United States production of 2,4-D and 2,4,5-T during 1967 and 1968, totaling about 14 million pounds. (Small packages for American homeowners were excepted.) These herbicides are effective against trees, brush, and broad-leaved plants.

The range of this scorched-earth policy, carried on with modern chemicals from the air, was of unprecedented magnitude.

Moreover, 2,4-D was mixed with picloram (which is the common name of the Dow Chemical Company's Tordon) to form a defoliant called "White" in Vietnam. Picloram, which is said to be the most toxic chemical toward plant life ever developed, is also extremely persistent in soils. Its persistence, in fact, is legendary. Several years ago mules were taken from an American pasture that had been treated with picloram and used to plow a tobacco field. The tobacco, growing, became stunted and deformed. Scientists discovered later that the picloram had passed unchanged through the mules' digestive system and blighted the tobacco crop.

In the face of rising concern within the American scientific community, DOD commissioned a study on the effects of its defoliation policies. The study was undertaken by the Midwest Research Institute, and one must marvel at the institute's valor. Because of the obvious impossibility of conducting such a study in the zone under fire, the institute made its study entirely in the United States on the basis of interviews and of research among existing studies on herbicides. The final report was issued early in 1968, after the National Academy of Sciences had looked it over and passed it along to DOD with the comment that it was a "creditable job." Yet this assessment was tempered by NAS president Dr. Frederick Seitz, who noted: "It is clear that the compilation of this report is only a first step in investigating further ecological effects of intensive use of herbicides."

The study came to the conclusion that the defoliation program carried on in Vietnam will cause no long-term damage there. "DEFOLIATION STUDY CASTS DOUBT ON LONG-TERM DAMAGE IN VIETNAM," proclaimed a headline in the *New York Times*. But anyone who took the trouble to read either the report itself or the newspaper articles based on the report, learned that ignorance, rather than fact, underlay that conclusion. It was based not on firsthand observation, but on the use of herbicides in this country's domestic agriculture and silviculture. Since neither the climate, nor the extent and methods of application in the two countries were in any degree similar, the conclusion could hardly be said to be based on fact.

The report admitted ignorance whether:

- The defoliation of the rain forest would drive a number of rare animals over the brink of extinction.
- The constant rain of herbicides would affect the water quality in nearby streams and lakes.
- The region's rainfall might be affected by the widespread defoliation.
- "The removal of forest cover may result in accelerating the conversion of [certain tropical soil types] into laterite rock, which would greatly and irreversibly lower the productive capacity of the treated ecosystem."

This report, by which DOD apparently intended to soothe the alarm of the scientific community, did not dwell on certain long-term effects of the herbicide treatment on the health of the Vietnamese people. Yet, almost simultaneous with the report's composition there arose speculation by scientists about the impact of serious alterations of the environment on "pre-scientific" populations. In a paper delivered to the 66th Annual Meeting of the American Anthropological Association, Dr. Alexander Alland, Jr., of Columbia University, recalled that the great plague which devastated England in the fourteenth century may have been triggered in part by the destruction of woodlands in order to grow crops.

"There is evidence, however, that disturbances of normal (meaning long term) ecological and/or behavioral patterns usually lead to increased disease rates and lowered viability . . . Genetic and behavioral adjustments to disease are part of a general process of adaptation to environment. Public health, particularly in pre-scientific populations, depends upon such accommodations."

This delicate balance was upset in the fourteenth century when wild rodents, harboring the plague bacillus, were driven out of the dwindling forests. Approaching settled areas, they passed on the bacillus (a flea served as vector) to highly susceptible domestic rats. After the death of these rats, the vector fleas sought other favorable hosts, which usually turned out to be human beings. There are similar reservoirs of the plague known to exist in Vietnam, and a drastic alteration of the forest could set in motion the familiar train of events.

The Department of Defense may, in carrying out its assignment, contribute to that flow of unsound knowledge that has formed the disquieting background of the pesticide controversy. At a time when scientists are only just becoming aware of the baffling long-term effects of pesticide use, the effort to understand them fully may suffer a setback in the interests of national policy. DOD does not seem to be aware that, even in the use of "harmless" herbicides, it is contributing to Vietnam's agony.

"We are too ignorant of the interplay of forces in ecological problems to know how far-reaching and how lasting will be the changes in ecology brought about by the widespread spraying of herbicides in Vietnam," writes Arthur W. Galston, Professor of Biology at Yale and president of the Botanical Society of America. "These changes may include immediate harm to people in the sprayed areas and may extend to serious and lasting damage to soil and agriculture, rendering more difficult South Vietnam's recovery from war, regardless of who is the 'victor.' "

Yet even a country which lives from day-to-day in the horror of war must feel the impact of DOD's herbicides. Thomas O. Perry of the Harvard University Forest pointed this out in a letter to *Science*.

"The long-term effects of spraying such an area may be imponderable," Perry wrote, "but the short-term effects of using these chemicals are certain: a lot of leaves, trees, rice plants, and other vegetation are dead or dying; and a lot of insects, birds, animals and a few human beings have either migrated or died of starvation. The North Vietnamese are fortunate — they have only bombs to contend with."

Prestigious societies joined the clamor for sanity. In the summer of 1968 the Board of Directors of the American Association for the Advancement of Science urged DOD to suspend the use of arsenicals used to kill rice and elephant grass in Vietnam until a thorough field study of their effects had been carried out there. Almost simultaneously, the Ecological Society of Japan resolved to demand that the United States immediately stop the large-scale military use of herbicides and forest burning in Vietnam.

The myth that herbicides can be poured indiscriminately into the environment with no ill effects was at last being destroyed.

18. The Two Faces of Public Health

IDEALLY, PEST CONTROL is practiced in a highly selective manner by the public health authorities. A public health team, fighting malaria in a tropical village, does not set out to eradicate the entire mosquito population of the area. Its target is the vector — an organism which carries and transmits the pathogen to man or his domestic animals. Accordingly, the team concentrates its fire on those mosquitoes resting on walls in the local houses. These are the insects, gorged on human blood, which carry the malarial pathogen (a sporozoan parasite) from one human being to another. In applying residual pesticides to selected internal surfaces of houses, the public health team attempts to bring the toxin into contact with the vector at a point which is most likely to interrupt the transmission of the pathogen.

Despite the tragedies resulting from pesticide use — even *approved* pesticide use — these chemicals have often worked wonders in parts of the world where public health advances have been slight. The Swiss chemist, Paul H. Mueller, who developed DDT as a pesticide, won the Nobel Prize for Physiology and Medicine in 1948. DDT had been spectacularly successful in combatting typhus and malaria in Italy during World War II, and it duplicated that performance later in other countries. Its side effects were not recognized then; nor was its effectiveness seen to be limited by the fact that target insects, like flies and mosquitoes, soon developed resistance to it (see *Silent Spring*, pp. 218-25), while the natural controls of those target insects were wiped out. Pest insects not only survived, they flourished.

Writes Robert L. Rudd in *Pesticides and the Living Landscape*:

> Greater numbers of a resistant insect population must derive not so much from an inherent advantage but from insecticide effects on non-target species normally competing with or preying upon the resistant form.
>
> Release from natural limitations of numbers has been repeatedly shown in insects of agricultural and medical importance . . . In Zanzibar, bedbug, flea, and chicken-mite populations increased following dieldrin spraying. In Southern Rhodesia, bedbugs resistant to BHC, DDT, and dieldrin increased to numbers not previously known in the area. In Phoenix, Arizona, resistant houseflies emerged in much greater numbers from dieldrin-treated privies; the reason seems to be the elimination of a highly susceptible competing species of fly whose presence normally renders privy contents unsuitable breeding places for houseflies. A. W. A. Brown* calls attention to increased housefly production in Liberia, Nepal, Saudi Arabia, Japan, Sicily, Sardinia, Egypt, Kenya, and Tanganyika following spraying. So the list grows; the problem is world-wide.

Yet, as Rudd points out, the magnitude of public health problems often dictates the use of chemical pesticides in those areas. By resorting to new chemicals, and sometimes stronger dosages, the public health authorities manage to keep on top of the malaria-carrying mosquito and other troublesome vectors. Such emergency measures clearly are not necessary in most countries of North America or northern Europe. The control of malaria in the United States, for instance, can be attributed to a number of factors besides the use of pesticides, among them public education, the elimination of breeding places for mosquitoes, the effective screening of houses, and the increase in the amount of federal aid for building. Public health spraying programs for vector control ultimately are worse than useless unless they are accompanied by social measures of this sort.

* "The spread of insect resistance in pest species." 1958. Advances Pest Control Research, *2*: 351-414.

One of the weaknesses of the U.S. Government's attempts to unravel and communicate to the public the nature of modern pesticide use shows up in the activities of the Public Health Service, which is a part of the Department of Health, Education and Welfare. On one level, PHS is intensely aware of the poisonous nature of chemical pesticides; it takes an active part in the diagnosis and treatment of acute pesticide poisoning, as well as in monitoring programs to detect their more subtle effects. Yet on another level PHS deals in the application of pesticides to control vector-borne diseases. Both of these activities, of course, fall under the heading of "public health."

PHS, then, is almost solely concerned with the *direct* effects of pesticides, for good or ill, on human beings. In its task of evaluating certain pesticide programs, PHS often comes to conclusions that are based on but one strand of a vastly complicated web. If a direct cause-and-effect relationship between a pesticide use (or misuse) and human health is not apparent, PHS sees no harm in it. But, as we have observed, the pesticide problem is ecological. The failure to see these programs in an ecological setting may produce either sudden spectacular disasters such as that which devastated the cats (and later the human population) of San Joaquin, or the veiled, long-term, environmental ills that are the chief concern of conservationists.

It is hoped that the various studies undertaken now by the Department of Health, Education and Welfare (of which both the Public Health Service and the Food and Drug Administration are subsidiary agencies) will contribute substantially to a better understanding of the extent to which pesticides have contaminated the environment. Though the Food and Drug Administration does not register pesticides (that is the duty of the Department of Agriculture), it advises USDA on whether or not the residues left on food by any given chemical may prove hazardous to human beings; FDA then sets the permissible levels of that pesticide on the various foods it is registered to protect from pests. The agency puts particular emphasis on its "total diet" study based on foods eaten by boys 16-19 years old — the growing boy who is traditionally our biggest eater. In general, FDA has found residues on various foods to be less than one per cent of the permissible levels.

Nevertheless, in 1968 FDA joined with USDA to recommend that DDT residues then permissible on thirty-six fruits and vegetables be drastically reduced. At that time, DDT residues of up to seven parts per million were allowed on some of these farm products. New residue maximums of 3.5 parts per million were suggested for such produce as avocados, carrots, cherries, and citrus fruits, while maximums of one part per million were suggested for such produce as beets, blackberries, cauliflower, peanuts, raspberries, strawberries, and turnips. Further reductions were being considered for future years because, according to FDA, most of these residues are higher than necessary for adequate pest control when DDT is used according to its label directions.

The key to the national pesticide monitoring program lies in the "community studies" undertaken by the Public Health Service. This project is akin to the PHS studies on smoking; it is not a laboratory project, but one in which scientists test a cross section of the populace for the effects of pesticide exposure. The cross section includes people who are continually exposed (pilots of spray planes, greenhouse workers, professional applicators, fanatical housewives who pick up the aerosol can at the buzz of a mosquito, etc.); intermediate groups (residents of towns that are continually sprayed in mosquito and Dutch elm disease programs or that are adjacent to heavily sprayed farmlands); and the general population. A variety of methods are used in "people monitoring." Healthy volunteers are monitored by needle biopsies, surgery patients donate "snips" of organs being removed, and autopsied cadavers are a source of adipose tissue and a variety of organs. It is agreed that the average American (who does not work either in the manufacture or application of pesticides) carries pesticide residues in the amount of about 12 parts per million in his fatty tissues. The PHS studies will provide more exact information.

Most of the Public Health Service's early work in this area was focused on acute toxicity. The older inorganic pesticides, such as the arsenicals, were highly toxic to human beings, and caused violent spasms leading to death. The new class of synthetic organic chemicals called organic phosphates (especially parathion) also bring on symptoms of poisoning almost immediately. But those other synthetic organic pesticides, the chlorinated hydrocarbons (DDT, aldrin, dieldrin, endrin, etc.), as we have seen, are much more devious in their effects. It is now fully understood by scientists that the new studies will require at least five, and perhaps ten years, to complete.

"You don't die from smoking a carton of cigarettes," one government scientist says. "In the same way, it isn't the acute aspects of pesticide poisoning which could be the most damaging. Today you just aren't sophisticated unless you're a chronic toxicologist. You've got to follow up."

And yet, despite this concern with pesticide residues in the environment, PHS sometimes contributes substantially to those residues during its vector-control programs. Consider its recent attempt to eradicate the *Aedes aegypti* mosquito from the United States, Puerto Rico, and the Virgin Islands. *Aedes aegypti*, the carrier of yellow fever, was brought to America along with its human hosts from Africa by the slave ships, aboard which it bred in the water tanks. It established itself in tropical American cities and towns, breeding chiefly in man-made water containers. The virus carried by *Aedes aegypti* found suitable hosts in both human beings and certain South American monkeys, and yellow fever became a New World scourge. It was finally brought under control in most of the New World early in this century (there has not been a case of yellow fever in this country in over sixty years); however, a strain called "jungle yellow fever," which is kept alive by monkeys and may be transmitted by mosquitoes to man, persists in some areas of Latin America.

Why, then, in 1964, did the Public Health Service set out on a campaign (at a cost estimated anywhere from $50 million to $100 million) to eradicate the *Aedes aegypti* mosquito in this country? PHS claimed that it was complying with a treaty entered into by the United States Government with the Pan American Health Organization. At a later date (after the spraying had begun) it became public knowledge that no such treaty existed. The program was based on a resolution of the Directing Council of the Pan American Health Organization in 1947, calling on the member nations to do their bit in combating any resurgence of yellow fever by eradicating *Aedes aegypti.*

During the 1960's, the United States Government came under some pressure from Latin American nations to get on with its share of the program. These nations had worked for years to root out and destroy *Aedes aegypti*, chiefly by sanitation methods. Now they expressed anxiety that *Aedes aegypti*, known to exist in the southern part of the United States (where it has done no harm because there is no yellow fever or dengue virus to transmit), might somehow be transported in a shipment of goods to Latin America. There, the offending mosquito might come into contact with a human being who had picked up a dose of yellow fever in the jungle. The possibility exists that the deadly cycle could begin once more. Apparently a few mosquitoes had been transported to El Salvador in a shipment of old automobile tires that had collected puddles of rainwater while stacked in a junkyard. The mosquitoes had been quickly disposed of, but the anxiety remained.

After World War II the Malaria Control in War Areas Program (MCWA), operating out of the Communicable Disease Center in Atlanta, Georgia, had been a major project of PHS. The program, however, became a casualty of the economy bloc in Congress during the Eisenhower administration. Now, with the *Aedes aegypti* alarm, the people at the Communicable Disease Center perceived a golden opportunity to recoup their former importance — and the funds that went with it.

PHS, therefore, instead of spraying outgoing shipments which might harbor *Aedes aegypti,* decided to mount a full-scale spraying program, using DDT, to eradicate the mosquito. Despite the failure of other widespread campaigns using chemical pesticides to eradicate an entire species (*i.e.* the fire ant), PHS assembled its program and appropriations without any opposition; PHS officials had only to whisper the word "epidemic," though there was little chance of one erupting in the United States, and everybody fell into line.

In fact, the major outcry was occasioned by the Public Health Service's determination *truly* to eradicate *Aedes aegypti.* When PHS tried to have every laboratory, public and private, in the country destroy its lab populations of *Aedes aegypti,* the scientific community rose up in arms. Chances of a mosquito escaping from a laboratory and founding a population in the wild were remote. This is especially true in the northern United States. Furthermore, *Aedes aegypti* is an invaluable experimental instrument in research. As one scientist said, "The destruction of the colonies would mean loss of 25 percent of the entire world research on mosquitoes, and at the same time render practically useless years of work and the vast foundation of knowledge already accumulated."

The Public Health Service, after listening to protests from the Entomological Society of America and other influential organizations, modified its plans in that direction.

The protest against PHS plans to spread DDT through the environment was not as successful. The people at the Communicable Disease Center had leaped into the project with both feet. As Congress provided the funds, PHS expanded the program so rapidly that retired entomologists were summoned out of retirement to fill the ranks.

"The demand for personnel was so great," one government

official has said, "that a lot of us considered many of the operating individuals grossly incompetent."

Among the most articulate opponents of the PHS project was Dr. L. A. Terzian, head of the Medical Entomology Division of the Naval Medical Research Institute at Bethesda, Maryland. Testifying before a House of Representatives Subcommittee dealing with HEW appropriations in 1965, Dr. Terzian described the habits of *Aedes aegypti*:

> *Aedes aegypti* is a tropical, city-bred, urban mosquito, and it is almost always associated with man. Since it is a town mosquito, and a rather poor flyer, it is normally never found at any great distance from human habitations. Dr. Fred Soper, the architect of the present eradication program, has himself noted that the flight of *aegypti* is usually limited to within 25 to 30 yards of its breeding places. Thus, it is pertinent to note that these characteristics alone prevent the rapid spread of *aegypti* to distant places even under comparatively optimal conditions. Its breeding places are almost entirely confined to artificial collections of water found in tin cans, rain-water barrels, cisterns, old tires, broken bottles, flower pots and other debris. In other words, in the junk found in those urban areas where sanitation is poor and living conditions at best are substandard.
>
> Thus, unlike the costly control or eradication programs required of other mosquito species which must be tracked down in ditches, rivers, ponds, swamps, in field, country, forest, or jungle, and which require ditching, draining, spraying, and other costly techniques and tremendous amounts of labor, the control and eradication of *aegypti* is a comparatively simple matter of cleanup and sanitation . . . In fact, any program of sanitation and improvement of slum areas, in line with the announced policy of the present [Johnson] administration in this matter, would take care, for the most part, of much of the cost of this specific project.

Another critic of the PHS program was conservationist Shirley A. Briggs. Writing in the *Atlantic Naturalist*, she pointed out that, with powerful vaccines and energetic sanitation, we had dealt effectively with *Aedes aegypti* from the building of the Panama Canal to the extensive tropical deployment of troops in World War II. DDT was not on the market at the time. Then Miss Briggs went on to say:

> In justifying the use of DDT, the Public Health spokesmen claimed that the eggs of *aegypti* are durable and might hatch later when moisture collected again. This disregards the basis of sanitation programs. No places are allowed to collect water long enough for hatching — about a week. Without a place in which to lay eggs, the insect simply does not go about dropping them at random, in hope of future puddles . . . It is, in fact, one of the insects most cooperative with the aims of a sound control program.
>
> With neat irony, it is also one least apt to be controlled by DDT. Very early in the use of DDT, the problem of insect immunity arose, mainly with mosquitoes and house flies. *Aegypti* is a very heterozygous insect, which gives it a population with many slight variations in genetic structures.* This is a reason for its value as a research animal, since some individuals can be found that are resistant to almost anything. Remarkable immunities can be developed easily. So, in the familiar pattern, DDT spraying on a large scale will simply develop immune strains faster. In fact, this has already happened in Jamaica, Haiti, and the Dominican Republic, as well as some other Caribbean islands.
>
> But public health officials, since the advent of DDT and its related chemicals, seem unable to think of other means of attack. Their reaction to developing immunity in the target insect is

* There is a common misunderstanding about the mechanism of insect resistance. An *individual* does not acquire resistance in response to one or more pesticide applications, as the zebra and giraffe in Kipling's story acquired stripes and blotches in response to the "stripy, speckly, patchy-blatchy shadows" of the forest. A few individuals contain in their genetic makeup the traits that cause them to be resistant to a specific chemical compound. By virtue of this "unnatural" selection they survive, passing on this trait through the genetic code to their offspring. A predominantly "resistant" population eventually comes into being.

apt to be a search for more poisonous chemicals, to which the same drawbacks will apply. Programs have been held up, seemingly in confusion, in new islands of immunity.

It was a further irony that the U.S. Public Health Service turned loose its spraymen in Florida, where neither DDT nor any other chlorinated hydrocarbon pesticide had been recommended for mosquito control by the state Board of Health for almost ten years. The Board of Health, in fact, was forced to administer the PHS program, the only one of its kind in the state.

There was an immediate public reaction. Writing in *Florida Audubon* for July, 1965, Irwin W. Fritz of the Florida Audubon Society reported that the sprayers disregarded orders to ask the permission of homeowners before spraying their property.

"The supervisor of the Palm Beach area and his crew moved on the populace with apparent police-state callousness," Fritz charged. "The sprayers were drenching the areas with DDT as if a certain number of tons of pesticides had to be dispensed before an approaching deadline."

And Shirley Briggs told of a letter from a Florida conservationist who explained that "the Public Health Service assertion of no damage to wildlife was because they did not look for such damage until months later, and of course made no check before spraying to compare populations . . . [There was the] all-too-familiar picture of incompetent and uninformed spraymen, lack of proper supervision, and brush-offs by public officials responsible."

The public outcry prompted some mitigation of the PHS program's damage. The solutions of DDT were reduced, other chemicals were substituted for DDT, the spraying was carried out only on property where mosquitoes actually were discovered, and no spraying was carried out on property whose owner objected to it. The U.S. Fish and Wildlife Service also took a greater part in the program, looking for wildlife losses and offering advice.

The program which never should have been authorized came at last to an inglorious end. Before leaving office, President Lyndon B. Johnson's budget proposals carried a notation that $905,000 should be set aside to carry on the *Aedes aegypti* program in 1970. The Nixon Administration, probing for soft spots in the budget, eliminated the program.

*

Today the World Health Organization and other public health groups are gaining in their battle against insect-borne diseases. Blending sanitary measures skillfully with pesticides, they are driving malaria and other scourges of mankind from such vast countries as India and the Soviet Union. The quantity of DDT used against malaria-carrying mosquitoes in India rose from 375 tons in 1952 to 21,000 tons in 1961, then dropped to 8,400 tons in 1964.

The battle is far from won. Strains of malignant malaria continue to frustrate medical science. Furthermore, at a recent WHO conference in Geneva, the organization admitted that the program to eradicate malaria in West Africa had bogged down because of widespread insect resistance to DDT. One official called it "a serious development, and underlines the urgency of the search for alternative means of effective control."

But the decreasing dependence of the public health authorities on chemical pesticides does not mean that the worldwide demand for them also is decreasing. As public health measures of all kinds continue to save lives, the world's population surges to unmanageable numbers. The additional hungry mouths to feed in the underdeveloped nations will soon require, according to agricultural experts, a *sixfold* increase in pesticide production.

19. "Biological Weaklings and Prosperous Ignoramuses"

RALPH NADER, a prominent critic of both industry and government agencies, has called the United States Department of Agriculture "the Department of Agribusiness." One does not have to be hostile to the department to recognize the aptness of the description. USDA's chief function is to serve a single group of American businessmen — the farmers. It is charged with other duties, of course, including service to the consumer, and so by implication to all American citizens. Its first obligation, however, is to the farmers and the farm bloc, and there the department finds the source of most of the pressures exerted upon it.

Modern technology has increased the difficulties that face USDA as it carries out its duties to bolster the nation's agriculture, just as this technology has complicated the task of all the other supervisory agencies of government. This is a part of our modern dilemma. USDA is not the first arm of the Federal Government to find itself frozen into a series of contradictory stances before it realized what had happened. This has been especially true of the department's attempts to come to grips with the pesticide problem. Briefly, let us try to place agriculture's troubles in perspective.

When this was a nation of small farmers, USDA's responsibilities and accomplishments were enormous. Through its research, financial assistance, technical advice, and influence on farm legislation, USDA played an indispensable role in the development of the most efficient (if sometimes destructive) agriculture the world has ever seen.

It is difficult sometimes, especially after reading the chemical industry's enthusiastic literature, to recall that the great upward curve of agricultural production in the United States began in the 1930's. DDT, at the time, was simply a formula on a piece of paper, tucked away in a dusty file. Many factors contributed to the increased yield of American farms — fertilizers, new machinery, water control, and improved seed lines. Farmers had pesticides at their disposal too. To solve their various pest problems, they turned to the arsenicals and fluorines, petroleum oils, lime sulfur, the dinitros, and such botanical compounds as red squill, rotenone, pyrethrum, and nicotine.* Here and there the farmers suffered losses but overall the yields increased, surpluses accumulated and crops were plowed under.

Meanwhile, hand in hand with technological innovation came changes of a subtler sort. By the end of World War II, farming had evolved from a way of life to an industry. The farmer acquired the approach and the philosophy of the manufacturer. The small family farm, which once had produced a variety of fruits, vegetables, and dairy foods, now merged with a dozen of its fellows to form a single vast "factory farm" producing a single crop. In the interests of economy, every move was directed, every machine was adapted, to plant, nourish, and harvest that crop — whether it was celery, apples, corn, or soy beans. And a new word — *monoculture* — found a place in the dictionary.

Monoculture is a word which causes the ecologist to shudder. The stablest natural community is a complex one. The vastly intricate web of the natural world establishes a system of checks and balances, usually suppressing outbreaks of a particular species and preventing crashes of another. Traditional ecolog-

* Red squill, a rat poison, is derived from a Mediterranean bulb; rotenone, an insecticide, from the South American cube root; and pyrethrum, also an insecticide, from certain chrysanthemums.

ical studies present a tropical rain forest as a stable community, in which an almost infinite variety of plants and animals flourish in that dynamic give-and-take we sometimes call "the balance of nature"; conversely, an Arctic tundra is considered an unstable community, where the absence of diversity periodically triggers spectacular crashes of animal populations.

"Stability in a biotic community," LaMont C. Cole of Cornell has said, "is to an important extent a function of the number of species present."

Monoculture, then, is a little like a bomb: it is created to make a profit, and it carries within it the ingredients for a disaster. When potatoes are planted in unrelieved monotony from here to the skyline, the diversity of insects and other organisms which once inhabited that landscape diminishes. Into the vacuum slide a small number of organisms (the Colorado potato beetle, the potato leafhopper, the potato tuberworm, the potato flea beetle, etc.) that share an affinity for potato plants. Like the fat boy in a wagon load of plump capons, the newcomers flourish. And, in flourishing, they cross that socioeconomic line that divides the anonymous insect from the pest.

"The population of an organism," says a report of the Ribicoff Subcommittee (1966), "increases to the economic threshold (creating a pest) because of some disturbance of the environment. Control measures mean another disturbance."

Cultivated plants, which are biological weaklings by their very nature, find no safety in unity. Monoculture only exposes them to a more concentrated attack. Populations of certain insects, previously held in check by the "resistance of the environment," explode. It is no wonder that the practitioners of the new monoculture reacted somewhat irrationally at the close of World War II to the introduction of DDT as the panacea for their self-created problems.

"Everybody went overboard on DDT," says S. C. Billings, chief of USDA's insecticide evaluation staff. "They forgot good agricultural practices. A farmer needs some chemical controls, but he needs other things too. In fly control, for instance, many farmers didn't even bother to carry away the manure. 'Let DDT do the job,' they said. If a farmer thinks chemicals can be sprayed on to cover up bad agricultural practices, he's licked before he starts."

That USDA also went overboard on the new pesticides is easily understood. Partly because of its own past accomplishments, the department found its power and responsibilities diminishing in comparison to those of several other government departments. The twentieth century had reached the farm. The successful farmer-businessman, with his vast acreage, college degree, and modern machinery, was less dependent on USDA than the poorly educated hayseed who had struggled with his scanty plot just a decade or two before.

USDA, then, reacted in the tradition of all bureaucracies which feel their position threatened by shrinking responsibilities. The department's impulse to fabricate programs which give it the illusion of "busyness" has been especially apparent in the field with which we are concerned here. It went into the business of promoting pesticides, springing to arms at the first whisper of a pest. But to brand an insect a pest when its numbers are too small to inconvenience man seriously is as faulty from the viewpoint of science as it is from that of linguistics.

Contributing to the haphazard planning and the overkill frequently observed in USDA pesticide programs is what seems to be a built-in scientific deficiency in the outlook of many of its employees. This deficiency has been summed up best by ecologist Frank Egler in discussing the approach of various scientific and technological disciplines toward the ecosystem:

> Agriculture is one of the most non-ecological of these fields, and agriculturists are some of the most non-ecological of these technicians. We must realize that agronomists are a one-species-oriented people. Their efforts are to plough, plant and harvest the greatest crops at the least cost, within one year or even several

months. In their training and in their thinking, they are strangers in time and space to the complex ecosystem. They want a simplified, if highly unstable environment.

Too many USDA officials characterize warnings about the indiscriminate use of pesticides as a threat to themselves as well as to American agriculture. (The publication of *Silent Spring* threw these people into a dither; they failed to see that its message was not the abolition of pesticides, but the gradual withdrawal of persistent chemicals and the integration of the others in planned programs with biological and cultural controls.) Too many USDA publications read as if they have been put together by the public relations arm of certain chemical companies. The department, in fact, has been one of the most effective in government in getting its story across to the public. Its Extension Service, in particular, has poured out to farmers a steady stream of advice and instruction that, in some cases, is considered of very dubious quality by conservationists (and wise agriculturists).

"Extension Services, more than any other forces, are responsible for this land-use picture," Ian McMillan, the California rancher-naturalist, wrote in *The Condor* in 1960. "What they have done, however innocently, is to encourage and advise toward maximum, immediate, economic exploitation of the land, without regard for end results. They have demonstrated and advocated only that which is most profitable economically. I have never noted any real concern for conservation of the future. Through their influence and tutelage we have, on the local level, a leadership of prosperous ignoramuses."

USDA has been insistent that applicators follow correct procedures in the use of pesticides. Correct procedures certainly were followed in its own large-scale Japanese beetle and fire ant programs, yet both resulted in ecological disasters. Only a great public outcry forced USDA to modify those programs. Initially, for instance, USDA began the campaign against the fire ant by bombarding it, and everything else that got in the way, with two pounds of heptachlor per acre over large areas of the Southeast. Rachel Carson, in *Silent Spring* (page 156), described the campaign as "based on gross exaggeration of the need for control, blunderingly launched without scientific knowledge of the dosage of poison required to destroy the target or its effects on other life." Under fire, the department reduced the dosage to one and a quarter pounds per acre, and then to one-quarter pound per acre, spaced at intervals of three to six months. Finally, in USDA's own words, "heptachlor was entirely replaced with the much less toxic mirex bait which is considerably less hazardous than heptachlor."

DDT and the flood of synthetic chemical pesticides which followed it were enormously successful for a time. It has been only since the publication of *Silent Spring*, for instance, that the decision-makers at USDA came to appreciate fully the implications of both insect resistance and pesticide residues. LaMont Cole has summed up the farmer's professional hangup:

> Modern farmers plant tremendous acreages to a single variety of crop plant in the belief that this practice increases efficiency because of the ease of planting, cultivating and harvesting. The modern trend is also toward clean cultivation, the elimination of hedgerows between fields, and the destruction of roadside brush. All of these practices reduce the diversity of species and assure that man will have to compete with opportunistic species for the harvest. If his effort to combat pests takes the form of applying a broad-spectrum pesticide, the diversity of species will be further reduced and the inherent stability of the system decreased. And if the pesticide is one that has a residual action so that the soil becomes toxic, the land will begin to approach the ecological status of a salt lake — an environment to which few species can adapt but where those that do succeed tend to build up tremendous populations provided that an ample food supply is present.

Problems abound. (Those created by pesticide residues are discussed elsewhere.) Rarely do pesticides eliminate the target population. A few pest insects survive, especially those which prove genetically resistant to the specific chemical compound in use.

The American Government's tendency not to be able to distinguish one collective state from another in the political world has been carried over by USDA into the natural world; all insects (unless, like bees, they have amply demonstrated their collaboration with us) are lumped as enemies. The toll is particularly high among predator insects. Predators, for a number of reasons, rarely are as resistant to chemicals as the plant pests: they are less numerous, thus statistically lowering their chances for survival; when their food (the pest species) becomes scarce, predators produce fewer young per litter/brood, thus lowering both their potential for increase and the odds of producing a resistant variant; and, finally, by consuming large numbers of already poisoned pest insects, predators accumulate massive doses of the fatal compound. Thus, the predators are killed off, removing an important check on the hard-hit pest populations, which now have a chance to build up once more to peak numbers. As Robert L. Rudd has said, "Most insect control leads neither to total removal of pest species nor to reduced necessity for its control."

Even our collaborators, insects vital to our economy, often are gunned down in the confusion. Organisms necessary to maintain healthy soils disappear under a rain of pesticides. Pollinating insects, on which most of our agricultural plants other than the grasses largely depend for their perpetuation, also succumb. Nothing so starkly illuminates USDA's self-defeating approach to pesticide use as its failure to protect the pollinating insects.

*

A great deal has been written about the hazards posed to wildlife by the increasing use of chemical pesticides in agriculture. The farmer or the bureaucrat who carries on his crusade against insects in the hope of increasing food production often discounts these hazards. He refers to his critics disparagingly as "bird lovers," or "mystics." Yet it is he who has buried his head in the sand. The loss of a bird to contamination hits at the foundation of all life. Ecologists know today that birds or lizards or whatever, which fall victim in great numbers to man's schemes for subduing nature, are symptoms of a diseased environment. Eventually, all of us will be affected.

But now we know something else that may be of more immediate interest to some people than even the quality or the duration of life. The destructive practices that kill song birds or ornamental fish also strike at our economy. It is well to remember that pollution of all sorts — whether it consists of domestic sewage, factory effluents, or chemical pesticides — costs the community heavily in dollars and cents.

Only lately has concern begun to spread about the hazards of pesticides to beneficial insects and thus to agriculture itself. Until now, the entomologist who was genuinely concerned about the indiscriminate destruction of insect life has had to depend chiefly on the bird watcher and the fisherman to lead the outcry against insecticide abuses. (Of three million insect species known to man, only 30,000 or .1 per cent are classed as pests.) Suddenly the carnage has become so widespread that the issue has overflowed from the ecological to the economic sphere. The agribusiness community, therefore, must increasingly divert part of its attention from the unalloyed benefits to the costs of pesticide use.

The destruction of beneficial insects — especially bees — by insecticides is not a wholly modern problem. Lead arsenate wiped out untold numbers of bee colonies pollinating apple trees because it was applied before the trees came into bloom. The development of dormant oils partially solved this problem by enabling the growers to delay the use of lead arsenate. But the exponential increase in the use of insecticides in recent years has increased beekeepers' problems to a similar degree. One example of the growing complexity of maintaining bee colonies in agricultural areas is the introduction of Sevin (a trade name for the carbamate insecticide, carbaryl).

Sevin, because it breaks down easily in the environment, is generally looked upon favorably by conservationists. Indeed, its substitution for persistent pesticides such as DDT has been widely applauded. Yet, precisely because it is not persistent, applications of Sevin must be repeated several times during the growing season, thus inevitably coming into contact with bees. Because Sevin has been used extensively in certain areas to control the gypsy moth, and in others to control earworms, it has proved to be a curse for beekeepers. The state of Washington is a typical example of the resultant agricultural dilemma. Catherine May, the congresswoman from Washington, brought the problem before the House of Representatives in 1968:

> In my own state, application of the chemical pesticide Sevin for control of ear worms on sweet corn is resulting in the poisoning of bees necessary for the pollination of many crops. Without the use of a pesticide, ear worm causes considerable damage to sweet corn, and at present, Sevin seems to be the only means available for control of this destructive pest . . . This situation has serious implications for American agriculture, Mr. Speaker, for a potential threat to crop production is evident from two directions. Without the use of this particular pesticide, one crop is damaged, but when the pesticide is used, an insect vital to other crops is destroyed.

There are about five million bee colonies in the United States. In 1967, according to Clarence L. Benson of the American Beekeeping Federation, ten per cent of these colonies were wiped out by pesticides. Over 70,000 colonies were destroyed in Arizona, 76,000 in California. A steady decline in the number of colonies has taken place in the last twenty years, despite the pressing needs of agriculture for bee pollination — needs that USDA officials estimate would require twenty times the number of existing colonies to satisfy.

USDA was slow to respond to the problem. In fact, the Interior Department led Agriculture in this area. In an order to his department during the summer of 1964, Interior Secretary Udall called for precautions to avoid injury to pollinating insects. Yet Agriculture Secretary Orville L. Freeman, in a pesticide policy statement at the end of that year, was much less specific about this question, although apicultural investigations were his department's responsibility.

USDA lately has begun to reevaluate some of its own pesticide programs (in the national forests, for instance) in the light of increasing bee destruction. But the use of insecticides by USDA personnel is slight compared to that of independent farmers. USDA unquestionably has been lax in its responsibilities in this broader area, as Clarence Benson pointed out in 1968 at a meeting between beekeepers and congressmen:

> Never in history has there been such a barrage of chemicals directed toward the destruction of the honey bee. Without question some will say this statement is not true. What then are the facts?
>
> Federal law requires that insecticides be registered and there are specific requirements for labeling. If labels are not in violation of the law there will be a warning on the label indicating that the material is toxic to bees. Various wording is used, but something such as the following is on the label: "Do not apply to crops in bloom." Consequently, no one could plead ignorance relative to the effect of those materials on bees . . . Contrary to such instructions, the federal government in publication after publication recommends the use of those materials to "crops in bloom."
>
> As an instance, Agricultural Handbook No. 331 on page 11 has this recommendation for cabbage looper: "Parathion on foliage as needed." Farmers Bulletin No. 2086 "Growing Pumpkins and Squash" devotes seven pages to the promotion of insecticides. In two small paragraphs there is some mention of bees. No instructions are given the grower relative to the pollination requirements of these crops in spite of the fact that it is known that these crops

must have insect pollination to produce *any* crop. It is also known that crops of pumpkins and squash have been produced without insecticides . . .

In addition to the federal government's recommending that growers of various crops use insecticides contrary to instructions on the label, the federal government itself engages in various programs using insecticides contrary to instructions on the label. USDA Plant Pest Control, for instance, in cooperation with the state and county in California, and with the state of Arizona, was responsible for the destruction of thousands of colonies of bees. In addition to this direct promotion and use by the federal government, federal funds are used by the various land grant colleges and state extension services to promote the use of insecticides contrary to instructions on the label.

In general, however, USDA recently has been inclined to review many of its pesticide programs. Like other federal agencies, it is reducing its dependence on the persistent pesticides. The U.S. Forest Service (a part of USDA) took the pledge conditionally in 1965.

"DDT's marketing problems began in earnest with the publication of the late Rachel Carson's *Silent Spring*," *Chemical Week* said in 1969. "Evidence: in '57 the U.S. Dept. of Agriculture sprayed 4.9 million acres with the pesticide; in '67, USDA sprayed only 100,000 acres with it. Last year the figure was zero."

USDA also has displayed a commendable willingness to cooperate with other federal agencies in reducing pesticide hazards. The regime of Orville L. Freeman under Presidents Kennedy and Johnson obviously took such cooperation seriously.

"Cooperation between Freeman and Udall warded off a collision between the two departments," an Interior Department official says, "and established a relationship no one would have thought possible only a few years before. The existing situation is not good enough — there is a lot to be done — but the groundwork has been laid that can be used when the pesticide controversy flares up again."

While USDA's responsibilities have shrunk with the farm population, it continues to exercise many important duties. These include the registration of pesticides marketed in the United States, and the supervision of the labeling that appears on pesticide containers. After the publication of *Silent Spring* and the resultant uproar, USDA established a more rigid system for reviewing all applications for the manufacture and marketing of pesticides.

Problems remain. Despite good intentions on the department's highest levels, remedies are slow to come about. The inclination, if not the entirely free hand, to overkill in the field persists, and the American landscape frequently receives doses of chemical sprays it does not really need. Capable secretaries like Freeman find it difficult to alter attitudes and policies that have gained a foothold among long-time employees. USDA men in the field are not likely to say to their farmer-friends, "Well, boys, it turns out I've been giving you bum dope for the last ten years!"

Late in 1968 USDA came under fire from the General Accounting Office for not acting aggressively enough to protect the public from "misbranded, adulterated, or unregistered" pesticides. GAO, which is the congressional watchdog agency, was reporting to Congress on USDA's enforcement of the Federal Insecticide, Fungicide and Rodenticide Act.

The report charged that USDA's Agricultural Research Service (1) had failed to remove some potentially harmful pesticides from the market; (2) had no procedures for determining when alleged violators of its regulations would be reported to the Department of Justice for prosecution; and (3) had not reported any of the violations it had uncovered during the last thirteen years. In 1967 alone, 1,147 violations were reported out of the 4,958 samples taken of pesticide products.

One case, for instance, involved a disinfectant sold to hospitals on the claim that it killed a bacterium (staphylococcus) that causes serious infections. Although a report in January, 1965, disclosed that the product was not reliable, USDA did not prosecute the offending company, and did not remove the product's registration for almost two years. Some months after its registration had been cancelled, the company was still shipping the product.

"We believe that cooperative action by a manufacturer in recalling defective or hazardous products is the most efficient and effective means of removing such products from channels of trade," R. J. Anderson, the acting administrator of the Agricultural Research Service, said in response to GAO's report.

The chemical industry has joined the farmer as a coddled client of the Department of Agriculture. It is difficult to root out loyalties forged by tradition and mutual backscratching. But if the concept of a Federal Committee on Pest Control is to have any validity, the various federal agencies, supported by firm standards of evaluation, must speak with the single voice required to solve our environmental difficulties. Until then, the regulatory apparatus designed to protect the public will continue to break down at critical points.

20. The Spray's the Thing

> *"I was wondering what the mouse trap was for," said Alice. "It isn't very likely there would be any mice on the horse's back."*
>
> *"Not very likely, perhaps," said the Knight; "but, if they* do *come, I don't choose to have them running all about."*

"I BELIEVE our greatest problem in the state is the policy of Iowa State University at Ames," writes Mrs. Darrell M. Hanna, a conservationist who lives in Sioux City, Iowa. She refers to a local campaign (eventually successful) in 1967 to establish a sound program for the control of Dutch elm disease.

"The University has actively promoted the use of DDT for Dutch elm disease all over the state and they have unlimited outlets for their propaganda," she continues. "They have come out with a brand new film which advocates the use of DDT and they are booking it all over the state. The worst part of it is that the people of our cities and towns believe and accept their advice."

The men who direct state agriculture departments, extension services, and agricultural colleges often share the White Knight's zeal for preparedness. These men offer advice throughout their state to anyone who wants to grow green things — from thousands of acres of corn to a small garden of roses. And their advice is *spray!* Be careful, of course, but *spray*. Spray before the plants emerge; spray at this stage of growth; spray at the first sign of a bug. Inevitably, the spraying conforms to the calendar, rather than to need.

Here, on the local level, where they most concern the majority of people, we are apt to find pest control programs at their worst. These programs usually are based not on scientific principles but on pork barrel politics and an aggressive sales pitch. Arboriculturists and professional spray applicators have been saturated by chemical industry literature, just as many doctors have been duped by the extravagant claims of drug manufacturers. The applicators, in turn, exert pressure on town managers before each "spraying season." Like patent medicine hawkers, they have a remedy for every ill. Town officials, in their turn, shop around to find the *most* spray for their money.

"The greatest advocates of promiscuous spraying are the boys coming out of school, and their knowledge is usually confined to what their professors told them," one city forester says. "I think that some of these professors must either be lazy or too close to the chemical companies to give their students proper views of the control of plant diseases and insects. I heard one young graduate tell a group that you did not have to know much about diseases and insects — just contact a chemical company and they would set up a program for you."

Indeed they will!

The programs thus conceived are carried out haphazardly, and with no notion of their ultimate effect on the environment.

Many communities spend thousands of dollars a year trying to eradicate mosquitoes with DDT, even when there is no suggestion of a health problem. In some cases, it is simply a matter of local citizens weighing a minor nuisance against a serious environmental hazard, and opting for the hazard; in most cases citizens are not convinced that the hazard is serious.

Sometimes economics, rather than science, restores a community to its senses. Beginning about 1956, the fishing for landlocked salmon drastically declined at Maine's famous Lake Sebago. The decline coincided with a considerable increase in the use of DDT by the owners of camps and resorts who had sought to protect their fishermen clients from insect bites.

A thorough study by state biologists confirmed both the high incidence of DDT residues in the salmon, and significant declines in the size and number of fish being caught. This news, coupled with the complaints of local merchants and resort owners that they had lost money since the onset of poor fishing, at last cleared the way for the abandonment of DDT near the lake.

Attempts to control the gypsy moth have occupied some state forestry departments excessively. This insect, which arrived in the United States a century ago, thrived in the absence of the predators that had usually kept it in check in its native Europe. All attempts to eradicate the gypsy moth have failed. F. J. Trembley, a biologist at Lehigh University, has described its naturalization in this country, where it has become a pest by feeding on the leaves of forest trees:

"Pestiferous insects have been moving over the face of the earth for a long, long time . . . Newly immigrant insects, if they survive at all in their new home, usually become extremely over-abundant for a few generations and then their numbers drop drastically. The new immigrant has been absorbed into the ecosystem. Pests like gypsy moths cannot continue to be as destructive to the trees on which they live as they at first seem to be and still survive."

Stephen Collins, associate professor of biology at Connecticut State College, has suggested why these "cosmetic" programs (often conceived because state officials fear that tourists' sensibilities might be offended by the sight of a partially defoliated landscape) are allowed to continue.

"I was employed by the state's Agricultural Experiment Station for five years," Collins says. "Many of the 'research programs' there were complete whitewashings of the true state of affairs. If evidence was uncovered that showed the pesticides were harmful, it was not released to the public. The conflicts in my work there — assessing the harm that these persistent pesticides caused and then seeing the report buried — actually made me physically ill."

There is a tendency to shrug off complaints against many spray programs with the remark that "the experts know best." Further investigation may undermine such comforting beliefs. In Maine, during a program which covered 100,000 acres with DDT for the control of spruce budworms in 1966, there was a farcical breakdown in communications. The State's Forest Commissioner justified the use of DDT partly on the basis of a report compiled on bird populations during an earlier DDT program.

This research, it was discovered after the 1966 program's completion, had been performed by a *high school senior*. The young man, who went on to study biology in college, told Aaron M. Bagg, president of the Wilson Ornithological Society, that he had "hoped the report would remain in obscurity forever. Unfortunately, this was the first effort I had ever made in the direction of scientific objectivity, and I feel that I was not entirely successful."

No single crusade against pests has raised passions so high, or has been waged with less justification, than the Dutch elm disease program. Since DDT came on the market at the end of World War II, it has been used in an attempt to stem the disease's advance by killing the beetles which carry the offending fungus from tree to tree. The air along elm-lined streets in old towns grew misty as pest-control crews laid down barrage after barrage of DDT. Still the disease advanced across the country. DDT, for Dutch elm disease control, often has been a dismal failure.

Some of the earliest indictments of that chemical and its long-term effects were the results of studies concentrated in areas where experts in the Dutch elm disease were at work. The sequence is familiar today: the leaves beneath the trees become contaminated by the spray, earthworms (which are highly resistant to DDT but store it in their tissues) eat the leaves, and robins in turn eat large numbers of contaminated worms. The affected area thus loses most of its robins.

Challenged, the advocates of DDT have simplified the issue to a choice between elms and robins. This, as Roland C. Clement of the National Audubon Society has written, is "childish nonsense." Clement went on to explain the basis of our problem:

"Much as we all love the graceful elm, it is first necessary to recognize that we planted too many of them: we have been guilty of 'putting all our eggs in one basket.' No sane investment counselor would advise such a lack of diversification. Diversification is nature's way of protection against catastrophe, just as it is in making up a portfolio of securities. We should therefore look upon the current loss of elms as an opportunity to right the balance by planting a better-diversified variety of shade trees (including scattered elms) because it is nearly certain that most elms will eventually succumb to this fungus disease."

Richard J. Campana of the Department of Botany and Plant Pathology at the University of Maine also has shed light on this aspect of the control program. "The concept of control is misunderstood by the public," he has said. "Spraying, even when properly done, protects only a limited number of elms. It is not possible to protect all elms in a community. Trees of high value such as those in parks and malls and around public buildings should receive the most attention. I do not recommend using DDT under any circumstances."

The soundest advice today is that which recommends sanitation (the removal of diseased trees and the burning of dead wood), or a combination of sanitation and chemical spraying. USDA and a number of state agencies no longer recommend DDT for use in Dutch elm disease control; some states, like Maine and Massachusetts, have banned it for this use. The most frequently mentioned alternative chemical is methoxychlor, a chlorinated hydrocarbon which, though quite toxic, has not the persistence that is characteristic of its siblings.

DDT is still used in these programs because it is cheaper than other chemicals. Moreover, even well-intentioned cities find that in the past they have tumbled to a huckster's pitch and bought immense quantities of DDT at "substantial savings." In that case, city officials feel that they cannot justify the "waste" of large stores of DDT still on hand. (New York City, plagued by this problem for a long time, finally banned DDT from its parks in 1969.) Washington, D.C., however, clings to DDT for no reason that conservationists can fathom. The results of indiscriminate spraying in the Washington area have been vividly described by Louis J. Halle, the author of *Spring in Washington*, who returned to that city for a visit in 1969 after ten years abroad:

> Rachel Carson's "silent spring" seemed already halfway to realization. Everywhere, the starling had replaced the robin as the common lawnbird. It has been remarked before that the gradual disappearance of a species, because it is a negative event, is less likely to be noticed than the arrival and increase of a new species. Among my Washington friends I found a less marked awareness than mine of what has been happening to our bird-life, because for them it has been happening gradually. But I was Rip van Winkle coming back to his old haunts, and what I saw presented itself as the catastrophe it is.

Even the Mall in front of the capitol has been heavily and consistently sprayed with DDT.

"Not only do they continue to use DDT," writes Shirley A. Briggs of the Audubon Naturalist Society, "but they do it at all the wrong times of year and weather. When a protest was made to the man in charge that he was spraying at the height of the warbler migration, he said that this made no difference because birds do not migrate in the city — only in the country!"

This piece of news must come as something of a surprise to experienced ornithologists and bird watchers, who believe that it is the warbler migration, and not the blooming of the cherry trees, which is the chief glory of Washington's spring.

*

The warning signals, in the form of sophisticated scientific evidence, have been posted for the enlightenment of state and local officials charged with the supervision of pesticide use. Yet too often the shoe is on the other foot; it should not be the burden of a citizen to assemble and present scientific information to a local government to correct serious abuses. Sound policies and information, ideally, should originate with the state and filter down to the local level. But are the states equipped to establish sound policies and disseminate sound information? To answer that question we must look more closely at the apparatus designed by the state to deal with pesticide problems.

A chief source of pesticide information locally is the State Extension Service. In 1966 USDA extended its Cooperative State Extension Service programs in pesticides to include all 50 states and Puerto Rico. Under this program, USDA assigns a Pesticide Coordinator to each state. The coordinators prepare educational materials and provide pesticide workshop training schools for farmers and other pesticide users in their states.

The quality of the "education" acquired by pesticide users under this program in Rhode Island may be evaluated from an article written by the state's Pesticide Coordinator for a Providence newspaper in 1967. The article appeared under the head: "BIRDS STILL SING AS PEST CONTROL PROGRAM GOES ON." After a reference to Rachel Carson's book, this official went on to prove that the spring of 1967 was hardly "silent" by numbering the various birds he had seen in his backyard. Because he was alive and happy, and was not tripping over the carcasses of dead birds on his heavily sprayed grounds, he felt that all was well in his antiseptic world.

This sort of reasoning, of course, completely overlooks both the vacuum which is quickly filled in nature by other birds taking the places of dead ones in areas where there seems to be an adequate food supply, and the sophisticated biological techniques required to detect those losses in the first place. (To follow the author's reasoning into the human world, one would have to conclude that, because a murdered man's apartment was taken over later by another tenant, there had been no murder at all.) The author concluded with a plea for "safe usage," which he seemed to feel may be summed up in wearing protective clothing while spraying, locking up pesticides when they are not being used, and burying the containers when they are empty.

Other states look to their universities for advice and information about pesticides. At the beginning of this chapter we noted the opposition of Iowa State University to conservationists who were struggling to ban the use of DDT for Dutch elm disease control in Sioux City. In Delaware, one occasionally hears pesticide advice and information discounted by conservationists because of the close relationship there between the University of Delaware and the world's largest chemical complex, E.I. du Pont de Nemours and Company. Nine of the fourteen trustees serving on the university's executive committee in 1968 were members of the du Pont family or were associated with the company.

"According to many faculty members and students," a reporter wrote recently in *Science*, "the university . . . has been 'distorted' and 'intimidated' by the du Pont presence . . . It is perhaps not surprising that a Delaware faculty member felt no qualms about publicly criticizing Rachel Carson's attacks on pesticides and the chemical industry, but when two faculty members asked permission to give testimony that was expected to be adverse to industry at a pollution hearing last year they were advised by the university administration to submit remarks in writing but not to testify in person."

However, Dale F. Bray, head of the University's Department of Entomology and Applied Ecology, is unaware of any pressure by du Pont upon his department. "If such exists, it is far too subtle for me and my colleagues to detect," he says.

As the pesticide problem balloons to more unmanageable dimensions, conservationists believe that the creation of state pesticide control boards generally is a step in the right direction. To date, only 32 of the 50 states have any sort of pesticide board. Most of these are simply advisory bodies. At the least, such boards can contribute information about pesticides to the public, and point out to various state agencies that there *is* a problem. At best, they can stimulate the cooperation and research that the Federal Committee on Pest Control has sometimes prompted.

Several states, among them Massachusetts, have granted their pesticide control boards certain regulatory powers. Usually these boards lead a sort of back-street existence like any other poor relation. The board is empowered to license both aerial applicators and those who apply pesticides "to land of another by other than aerial means," the so-called custom applicators. Licenses are granted on the basis of examinations prepared and given by the board. Composed of the state's commissioners of agriculture, natural resources, and public works, and the chairman of the state reclamation board, the pesticide board also is charged with studying current pesticide research and making the results of such studies known to various state agencies and the public.

"About all the board accomplishes is to keep spray planes from colliding over a town," says Wayne Hanley of the Massachusetts Audubon Society. Of an examination given to professional applicators which he had observed, Hanley comments: "Most of those people work in town sewage plants during the week and spray trees on Saturdays. At the exams they stand around in little clusters, peeking over each other's shoulders to try to find the answers to the questions. The worst part of it is that most of them can't do the arithmetic needed to figure out the proper proportions when they mix the stuff."

The Maine State Legislature established a pesticide control board, but did not appropriate any funds for its operation. In Connecticut, a vice-president of a local utility company, notorious for its herbicidal "brownouts" along utilities' rights-of-way, was appointed chairman of the pesticide control board. New York became the first state to appoint an ecologist to its board; the ecologist, George M. Woodwell, was given simply advisory status.

"But the advisors will point the direction, if any, in which the board moves," Woodwell said. "The majority of any such board is made up of men in state administrative positions, and they are always politically vulnerable. They have a vested interest and do not want to rock the boat, or make trouble for the administration." Here Woodwell gave a little shrug. "Remember, this is not really a pesticide *control* board. It is an advisory board, and thus it is mainly ineffectual."

Ideally, a state pesticide control board should propose its own legislation. It should go beyond the investigation of problems, and seek to modify state programs, regulations and policies on the basis of its findings. Only when it advises the legislature of the state's needs, and the legislators themselves come to it for advice, will pesticide control boards begin to serve the purpose for which they were created.

*

Here and there we find bright spots as a state pursues a sane pesticide program. For some years now, California's agencies, especially its Department of Agriculture and its Department of Fish and Game, have followed a program that should be a model for other states.

"California has, I think, always been the leader in regulation and surveillance," Robert L. Rudd of the University of California said recently. "This is not a provincial statement. We have the most acceptable set of advisors and regulatory devices anywhere, including those in the federal machinery. Almost all states pattern their regulations after California, with models from New York State a close second. I see only one major difficulty: the awareness of the true magnitude of the pesticide problem. (The same is true of land use and urbanization, though it is not true of water use.)"

California, of course, uses enormous quantities of pesticides. Its agriculture is at once huge, intensified, and ever-changing. In one recent year, more than 13 million acre-treatments were made with pesticides (about 6 million acres in agriculture and forestry were treated an average of 2.2 times). It is estimated that about 1/15 of all the land under cultivation in the United States lies within California's boundaries, while 1/5 of the national pesticide use is applied there.

In 1963, alarmed by fish kills and other environmental disasters, Governor Edmund G. Brown established a permanent committee on pest control methods and pesticidal effects. The committee was placed under the State Administrator of Natural Resources. In 1964 the California Department of Agriculture restricted use of many of the persistent pesticides by requiring a permit before they can be applied to fields. It also eliminated the practice of treating rice seed with DDT before planting in an effort to reduce residues found in pheasants. A state-wide hotline keeps the Agriculture and Fish and Game people aware of pesticide problems, and a report of wildlife losses will trigger an immediate investigation.

When one of the chemicals listed as injurious material by the California Department of Agriculture is used, the state requires a prominent DO NOT ENTER sign to be posted at all normal entrances to the land. This warning poster lists the pesticide used, the date of application, and tells "all persons . . . to stay out for two weeks."

California has been a pioneer in using integrated controls. In *Pesticides and the Living Landscape*, Rudd has described the campaign against the spotted alfalfa aphid (*Therioaphis maculata*) in California's alfalfa fields. Chemical pesticides proved disastrous, since the aphid's predators were wiped out. After some initial success in controlling the aphids with parathion and malathion, the population had swollen to catastrophic numbers, threatening the entire alfalfa crop.

"A return had to be made to a better degree of natural control," Rudd wrote. "In many trials, a systemic insecticide (Systox) reduced aphid populations, while harming natural enemies very little. Moreover, insecticide-induced mortality did not have to approach 100 percent for satisfactory control. Lower doses could be used. Meanwhile new enemies — insect parasites and fungi — were successfully introduced. The combined actions of a selective insecticide and natural enemies, both native and introduced, now keep the aphid populations in satisfactory check without excessive cost, great toxic hazard, or development of insecticide-resistant populations."

About twenty insect and weed pests have been controlled solely by biological means in California since 1923. Another thirty insects, mites and weeds are presently the objects of further studies in biological control. The 1965 Report of the President's Science Advisory Committee (*Restoring the Quality of Our Environment*, pp. 267-8) reported on one such control program in California that has proved to be both a biological and a financial success:

> The Fillmore Citrus Protective District was initiated in 1922; a number of citrus growers banded together to eradicate the California red scale by chemical methods. From 1922 until the early '50's fumigation with cyanide, and oil sprays, were used. A parathion program was continued until 1961, when the District switched to a total program of biological control. Parasites and predators were reared in the District's private insectary by an entomologist and technician on the District's payroll. In 1964, this insectary reared and released 10 million *Metaphycus helvolus*

against the black scale and 13 million *Aphytis melinus* against the red scale.

In 1964, the District had 331 growers with about 8,000 acres of citrus. The assessment to members was $24 per acre in 1960, the last year of the chemical program. In 1961, the figure fell dramatically to $6 per acre and in 1962 and 1963, to $8 per acre. The annual savings derived from the biological control program are estimated at $300,000 for the District, or about $40 per acre. The average cost of citrus pest control in the District was about $20 per acre in 1963-64, including the assessment. In other areas this figure was as high as $50 per acre for mites alone, and probably averaged $80-100 statewide.

But all is not well in the state agricultural structure. Both the College of Agriculture of the University of California (which is, itself, a land grant college) and the Agricultural Experiment Station are in trouble. The more immediate aspect this difficulty wears is financial. Substantial cuts in both research and extension funds are being made in the state legislature. But there are more subtle difficulties at the university too, difficulties which have their roots in a changing world and which have been widely publicized in their manifestations at Berkeley.

"The main issue is not really financial but re-definition of social roles," Rudd said recently. "Involvement is the central university issue now as always; but with whom? Do we serve producers and distributors or do we, in addition, serve consumers, farm workers, and, most pointedly, minority groups? We know of departments of agricultural economics, animal husbandry, and so on. But why no agricultural sociology, no human husbandry? Certainly the traditional role of the agricultural phases of the Land Grant College System is rapidly changing. Rachel Carson and I happened to hit hard early on one facet of needed change."

Part V

A Light at the End of the Road

21. Toward a Court of Last Resort

"We have subjected enormous numbers of people to contact with these poisons, without their consent and without their knowledge," Rachel Carson wrote in *Silent Spring*. "If the Bill of Rights contains no guarantees that a citizen shall be secure against lethal poisons distributed either by private individuals or by public officials, it is surely only because our forefathers, despite their considerable wisdom and foresight, could conceive of no such problem."

In recent years a vocal and knowledgeable lawyer named Victor J. Yannacone, Jr., abetted by a group of intense young scientists, set out to remedy this oversight on the part of our founding fathers. Yannacone regards polluters with unalloyed loathing, and run-of-the-mill conservationists with an ill-concealed scorn that is tempered only by pity. At first even stouthearted conservationists were put off by his militancy. Eventually, however, Yannacone's successful rushes at the powerful chemical industry convinced even the most squeamish that the best hope for progress in the struggle against the persistent pesticides might lie after all in responding to his irreverent battlecry: "Sue the bastards!"

*

Strong actions, as well as strong words, are long overdue. For too long the citizen who claims a clean and healthy environment as his consitutional right has faced a legislative and administrative stone wall. He parades his rights in petitions, public meetings, and letters of protest. The polluter marshals his power behind the scenes. It is an unequal struggle.

The case of the citizen whose rights are obscured by confusing social traditions is not without precedent, of course. The American Civil Liberties Union and the Legal Defense Fund of the National Association for the Advancement of Colored People have written a long history of successful court action in these matters. Much of the progress of American jurisprudence in the fields of civil rights and human rights springs from this litigation. There has been no comparable force in the field of environmental pollution. Yannacone and his colleagues organized the Environmental Defense Fund to fill this vacuum.

"A court of equity is the only forum in which a full inquiry into questions of environmental significance can be carried on," Yannacone says. "Only on the witness stand, protected by the rules of evidence though subject to cross-examination, can a scientist be free of the harassment of legislators seeking re-election or higher political office; free from the glare of controversy-seeking news media; free from the unsubstantiated attacks of self-styled experts representing vested economic interests and yet who are not subject to cross-examination."

Once again, it was a woman who took the initiative. In the spring of 1966 Mrs. Carol Yannacone of Patchogue, Long Island, learned of a fish kill that had taken place at nearby Yaphank Lake. The kill was attributed to the use of DDT in mosquito control operations. As a girl, Mrs. Yannacone had known the area well, playing on the lake's shores and swimming in its waters. Most people would have shrugged off the incident as simply one more manifestation of "progress."

But Mrs. Yannacone continued to be troubled by the contamination of Yaphank Lake. She talked it over with her husband, Victor, who already had had some experience in pollution litigation, and saw in the present case an opportunity to strike a blow against the pollution that seems to be closing in on all of us. The next morning he filed suit to restrain the Suffolk County Mosquito Control Commission from using DDT.

Yannacone was aware that he needed more than good intentions to make his case stand up in court. Fortunately, competent scientists abounded in the neighborhood. Among the nearby institutions were the New York State University at Stony Brook and the Atomic Energy Commission's Brookhaven National Laboratory. A number of scientists agreed to help Yannacone. It was their intention from the start to restrict their group to scientists. (The polluters in these cases like to characterize conservationists as "mystics" and "sentimental birdwatchers," but it is difficult to ridicule scientists who come to court supported by a string of degrees and a large body of research in their specialties.) On the basis of affidavits submitted by his new colleagues, Yannacone obtained a temporary injunction to stop the Commission from applying DDT.

Lawyers generally prefer to settle their cases out of court. Yannacone approached the matter in a different light. Though American law traditionally is antagonistic to the idea of litigation by private citizens against public bodies, he decided for two reasons not to compromise:

1) In cases which involve a basic principle, a trial is desirable so that scientific evidence can be put on the record.

2) Those who want to go on contaminating the environment must be made to stand up in court and tell *why!* Yannacone's method is public exposure. It embarrasses the polluter while educating the public.

Fittingly enough, the trial was financed in part by the Rachel Carson Memorial Fund of the National Audubon Society, which had been created after her death by some of her friends and admirers. Yannacone and the scientists who took part in the action against the Mosquito Control Commission gained valuable experience, since there was almost no legal precedent from which to draw. They came to the conclusion that legal action to protect the environment from polluters should be based neither on economic nor aesthetic grounds. The plaintiff must establish by sound scientific evidence that the offending program will cause "serious, permanent and irreparable damage" to an essential or a unique natural resource belonging to *all* the people of the community. Only on this basis is it possible, they believe, to get injunctive relief.

The county did not really defend itself; county officials claimed they were simply doing their duty, under their grant of police powers from the state, in controlling mosquitoes. The Mosquito Control Commission then revealed that it was phasing out the use of DDT, but still wished to use it in certain areas. The minutes of a recent commission meeting, brought out in court, indicated considerable confusion within the commission itself.

"Listening to everything we are going to do, we are admitting that misuse of DDT can be dangerous," one commission member said at the meeting. "This morning is the first time I have heard it stated that DDT is wrong."

In other words, the court action was educating even the Suffolk County Mosquito Control Commission!

The commission left its defense up to the state of New York. The defense was spotty, misleading — and sometimes revealing. The Department of Agriculture and Markets based part of its defense on DDT's alleged prowess in obliterating equine encephalitis, stressing the importance of horses to the state because "revenues from pari-mutuel betting are of vital necessity to the support of State Government." Meanwhile, as Roland C. Clement of the National Audubon Society pointed out, no one apparently had told the Department of Agriculture and Markets that "an inexpensive vaccine will amply protect the horses on which New York State's budget seems so dependent."

A Memorandum *amicus curiae* by the New York Department of Health set up many straw men, one of them being the attribution to the plaintiff of a claim that "DDT is solely responsible for the disappearance of the osprey." The plaintiff had made no such claim. Yet the Health Department went on with the counterclaim that the U.S. Fish and Wildlife Service now believed that "the decline of the osprey is a result of human activity and encroachment upon its breeding grounds." This ploy was disposed of when the plaintiff secured a vigorous denial from the Director of the Bureau of Sport Fisheries and Wildlife; the bureau had never said any such thing.

During the trial it was revealed that, *ten years before*, a high official had recommended that the state not use DDT in its gypsy moth control programs. This recommendation was made by James E. Dewey, who was responsible for coordinating the pesticide program of the State College of Agriculture at Cornell, to the Department of Conservation. The recommendation was buried in a state file. Despite the interest which this document might have had for many people during those years, its existence did not come to light until Dewey was called under oath to the witness stand by Yannacone.

Yannacone lost the battle but won the war. At the end of the trial the judge ruled that the prohibition of DDT is a matter not for the courts but for the state legislature to decide. Nevertheless, DDT has not been used since by the Suffolk County Mosquito Control Commission. More importantly, despite the adverse decision in court, Yannacone and the scientists had snatched the victory by convincing the public through news media that its case was just. The resulting public pressure brought about the policy changes that Yannacone went to court for in the first place. Finally, it is not probable that revelations and corrections of a fundamental nature would have been made during legislative or administrative hearings; nor would the scientists' voluminous technical testimony about the action of pesticides in the environment (its worth proven under vigorous cross-examination) have been put on the record where it is now available to other victimized citizens.

But the struggle had just begun. Reluctant to disband their effective group, Yannacone and the scientists incorporated instead, forming the Environmental Defense Fund (EDF). Yannacone became the Fund's chief counsel ("We're often fundless," he says mournfully) and Stony Brook's Charles F. Wurster, Jr., became the head of its scientists' advisory committee. Wurster is an idea man for the position. He is as militant, if not as voluble, as Yannacone, and he is an indefatigable correspondent. Largely through his efforts EDF has assembled over 100 leading scientists who are ready to stand up in court and support EDF's environmental crusade with their knowledge and reputations. Wurster is EDF's leading spokesman on the witness stand because of his extensive background in research, both in the laboratory and in the field, on the effects of persistent pesticides.

"Charlie is very, very bright and remembers everything he reads," George M. Woodwell of the Brookhaven National Laboratory told *Science*. "I think, unquestionably, he knows more about persistent pesticides than anybody in the world."

EDF has followed up its eastern success. In 1967 it filed suit in western Michigan to restrain nine municipalities from using DDT for Dutch elm disease control. Again, EDF lost in court but attained its objective. The Cooperative Extension Service of Michigan State University withdrew its statewide recommendation of DDT for this program and recommended instead sanitation methods coupled with the supplementary use of methoxychlor.

Encouraged by this success, EDF expanded its court action to include another 47 Michigan municipalities. By 1968, 50 of the 56 municipalities planning to use DDT had consented to the court orders which compelled them to use alternate methods of control. At the same time, an EDF suit temporarily averted a planned application of three tons of dieldrin over Michigan's Berrien County for the control of Japanese beetles. Wurster testified that from 10 to 80 birds and mammals, up to the size of cats and even sheep, would be killed for every beetle killed in this program. Though the Michigan and United States Departments of Agriculture eventually went ahead and sprayed 3,000 acres with dieldrin for what they admitted amounted to only about a single beetle per acre, EDF moved the country closer to a sane pesticide policy. The widespread publicity given both court actions publicized the strong scientific case against the persistent pesticides and publicly discredited the spray programs.

Events moved swiftly in the Great Lakes area after that. When high DDT residues and heavy mortality among fry threatened its expensive coho salmon program for sports fishermen, the Michigan Conservation Department aggressively condemned the use of the persistent pesticides. A further uproar occurred when the levels of DDT found in fish (up to 19 parts per million in salmon seized by the state) threatened to close down Michigan's commercial fisheries. In April, 1969, Michigan became the first state in the union to outlaw the sale of DDT within its borders.

By that time, EDF had already carried the fight into Wisconsin. The battle lines were drawn up when the Wisconsin Department of Natural Resources decided to hold an administrative hearing on the effects of DDT in the environment. EDF, supported financially and morally by the National Audubon Society's Rachel Carson Memorial Fund, the Izaak Walton League, and the Citizens Natural Resources Association, unveiled the latest worldwide evidence against DDT. The chemical industry, sensing that DDT was on trial for its life, formed a defensive "task force." It was composed of Allied Chemical Corporation, Diamond Shamrock Corporation, Olin Mathieson Chemical Corporation, Lebanon Chemical Corporation, Montrose Chemical Corporation of California (the nation's largest manufacturer of DDT, which is jointly owned by Chris-Craft

Industries and Stauffer Chemical Company), and the Geigy Chemical Corporation. All of these companies manufacture DDT except Geigy, which developed the pesticide in its plant in Switzerland.

At the start of the hearing the "task force" chose to have itself represented by Louis A. McLean, the Velsicol spokesman who has figured so prominently in attacking those scientists who have questioned the unrestricted use of pesticides (see pages 49, 263). Yannacone promptly put his opponent on the stand as an EDF witness. He forced him to sit there while Yannacone read portions of McLean's celebrated article in *BioScience* accusing the "antipesticide" people of being "preoccupied with the subject of sexual potency." Later McLean blandly suggested that EDF's scientific witnesses were not qualified to discuss the effects of pesticides on wildlife.

Personalities, however, faded into the background as headlines once more told the world what it means to inject DDT into the open environment. Yannacone and his colleagues were gaining ground on all fronts. The success of EDF's double-pronged attack in both the courts and the press may be measured by the growing response from the agri-chemical trade publications. In 1968 *Agrichemical West* wondered why EDF enjoyed a "tax free status" (while not mentioning the many tax loopholes available to the chemical industry's lobbyists) and suggested that readers complain to their congressmen. "It is bad enough to have a bunch of nuts witch hunting in your pantry without paying them to hold up your food supply!"

This high level of discussion was repeated in *Farm Chemicals*, which set forth the names and addresses of EDF's Board of Trustees in a two-column box accompanying its article on the organization. "Some of the names of the Board members should be familiar to you," the publication told its readers in prose pregnant with implication. Later it presented an editorial in which it added EDF to a list of "cults" threatening the existence of the chemical pesticide industry.

"Scientists themselves literally ostracized Rachel Carson," the editorial said, "and they will have to come to grips with this eroding force within their ranks."

The strident note was reminiscent of the outcry in 1962.

22. Toward a Complex Community

In 1964 a Harvard biologist named Carroll M. Williams welcomed to his laboratory Karel Sláma, a young entomologist who had come to the university from Czechoslovakia on a postdoctoral fellowship. Sláma brought with him over 1200 larvae of the linden-bug, a harmless European insect that lives on the fallen seeds of the linden tree. Both men were interested in the function of certain insect hormones. Sláma, especially, was anxious to clear up some questions about those hormones and his linden-bugs, which he had raised successfully for years in a Prague laboratory.

The two men, having stored the larvae in incubators, waited for them to complete metamorphosis and reach sexual maturity. For a while all went as expected. The larvae passed through the customary five larval moults which nature has programmed into the linden-bug's life cycle. Then the unexpected occurred. Instead of metamorphosing into winged adults, the 1200 larvae (with a single exception) underwent a sixth, and sometimes a seventh, moult. Those that passed through a seventh moult reached what, for a linden-bug, is gigantic size. Invariably, since the larvae were unable to complete metamorphosis, the process ended in premature death.

Both men were mystified. The young Czech assured Williams that nothing like this had ever taken place in Prague. Sláma sent home for another supply of larvae, which he carefully tended in his own quarters at Harvard; yet the ill-starred insects met a fate identical with that of their predecessors.

Fortunately, Williams is one of the world's foremost authorities on the part played by hormones in the process of metamorphosis. There are three separate internal secretions, or hormones, which regulate insect growth and metamorphosis from larva to pupa to adult. One of these secretions is called juvenile hormone. It is secreted by two tiny glands located in the head. At certain stages in the insect's life the juvenile hormone must be present, while at other stages it must be absent if the insect is to develop normally.

"An immature larva has an absolute requirement for juvenile hormone if it is to pass through the usual larval stages," Williams says. "Then, in order for a mature larva to metamorphose into a sexually mature adult, the flow of hormone must stop . . . The periods when the hormone must be absent are the Achilles' heel of insects. If the eggs or the insects come into contact with the hormone at these times, the hormone readily enters them and provokes a lethal derangement of further development. The result is that the eggs fail to hatch or the immature insects die without reproducing."

Williams, during his studies on insect hormones in the 1950's, had grasped the significance of these facts in their application to pest control. If their own juvenile hormone could be used against pests at a certain stage of their development, the two major disadvantages of modern pesticides — their harmful effects on non-target organisms and the development of insect resistance to them — would be circumvented. (An insect population could not develop resistance to its own growth-regulating hormones without committing suicide.)

But there were barriers to the commercial application of these facts: since it is impractical to extract juvenile hormone in great quantities from the insects themselves, the hormone would have to be synthesized by industrial processes. Moreover, although it would not be likely to act on any creatures except insects, the synthetic hormone also would have to be restricted in its effect to the target species; otherwise, bees and other beneficial insects would be wiped out along with the pests.

Williams, then, rightly diagnosed the destruction of the captive linden-bugs at Harvard to be the result of their contact with a source of juvenile hormone. He and Sláma carefully examined the contents of the incubators. Perhaps the linden seeds or the water vials, perhaps even the petri dishes in which the larvae were reared, had become contaminated by juvenile hormone under study elsewhere in the laboratory. But tests on all of these objects proved negative. Then the two men turned to the strips of paper towelling which had been placed in each dish to increase the surface available to the bugs.

"In Prague, filter paper had been routinely used in the cultures," Williams says. "We were astonished to find that when the towelling was replaced by a corresponding fragment of filter paper, the entire phenomenon vanished and all individuals developed normally."

A whole new range of experiments now suggested itself to Williams. The larvae were placed in contact with other kinds of paper. Whenever the paper was of European or Japanese origin, the larvae completed metamorphosis; whenever the paper came from North America, the larvae ballooned to "monsterhood." Kleenex, Doeskin Facial Tissues, Scotkins Paper Napkins, and Cut-Rite Wax Paper all showed high juvenile hormone activity. When Williams used pages from the *New York Times*, *Wall Street Journal*, *Science*, and *Scientific American*, he found them even more active than the towelling and napkins. The *London Times*, however, was "completely inert." Japanese newspapers caused a momentary flurry of excitement in the laboratory by proving as active as their American counterparts, but this development fell into place when Williams learned that the Japanese print their newspapers on Canadian newsprint.

The signposts had been uncovered. Further study traced what Williams calls the "paper factor" — the substance that activates the juvenile hormone — to the balsam fir, a North American evergreen of exceedingly ancient lineage which supplies much of the pulp for paper manufacturing in the United States and Canada. The tree's ancestors preceded insects on the earth. Since they are pollinated by the wind, balsams flourish independently of insects and in fact have synthesized certain terpenes which serve to keep them remarkably free of most pests. (Exceptions include the balsam-fir sawfly and the spruce budworm.) Perhaps at one time the balsams were obliged to protect themselves against a relative of the linden-bug, which belongs to the Pyrrhocoridae, an insect group also comprising the "red cotton bugs" which are especially destructive to cotton crops in India and North Africa.

"The most intriguing possibility," Williams says, "is that the paper factor is a biochemical memento of the juvenile hormone of a former natural enemy of the tree — a pyrrhocorid predator that, for obvious reasons, is either extinct or has learned to avoid the balsam fir."

Williams and Sláma, with the unwitting aid of their linden-bugs, had discovered in paper products a massive supply of an insecticide whose action is restricted to one group of insects. Here is the specific action that critics of pest control techniques have called for during the last two decades. Other terpenes, restricted in their action to small insect groups, have been discovered since in the balsam fir as well as in several additional species of North American evergreens. Japanese scientists have found similar terpenes in yews, while European scientists have found them in a number of primitive plants, including fiddlehead ferns (long the source of a folk remedy for parasitic worms). Synthetic juvenile hormone, recently prepared from terpenoids by Williams and his colleagues, has blocked the emergence of adult *Aedes aegypti* mosquitoes, which are vectors of yellow fever, and human body lice, which are vectors of epidemic typhus. The substance also blocks the embryonic development of their eggs. Williams, in fact, believes that synthetic juvenile hormone eventually will prove to be most effective as an ovicide.

Here is one example of that "other road" which Rachel Carson called for in *Silent Spring.* Man, in turning to biological controls, uses an intelligent method to deal with pest species. He encourages a complex community rather than a simpler (and therefore a less stable) one. In general, biological and chemical controls, directed against specific pest organisms, are not yet feasible on a large scale; yet USDA, industry and the universities carry on their research to develop:

• Techniques for sterilizing male insects with chemicals, irradiation or hybridization. During one recent program in California, USDA planes dropped 250,000 sterilized pink bollworm moths daily on cotton fields in the lower San Joaquin Valley. When the male moths, sterilized by exposure to cobalt-60 irradiation, mate with females in the area, they crowd out the normal males and prevent reproduction in the local population. "If this is successful," a USDA official says "it would reduce the need for ballworm spraying by 75 per cent." These techniques are especially effective when the pest population already is limited either in size or in area. Here the insects' own powerful sexual drives cause them to administer the *coup de grace* to their struggling population.

• Techniques for genetic manipulation, in which insects, carrying some fatal flaw in their genes, are released to pass on the defect to others of their kind (just as Queen Victoria's female offspring and descendants carried hemophilia into royal bloodlines on the continent).

• Sexual attractants, which confuse insects during their mating season, causing them to lose the trail of females; or lead them to traps and poisoned baits. In one instance, USDA researchers have isolated the chemical in human sweat which attracts *Aedes aegypti* mosquitoes. Scientists believe that this chemical, L-lactic acid, might be synthesized to lure mosquitoes to traps rather than to human skin.

• Methods for spreading disease among insect populations. One of the most successful disease agents has been *Bacillus thuringiensis*, a bacterium which attacks and kills specific insects.

• Control programs in which predatory insects or parasites are released to prey on pest species. In 1967 New York State released tiny flies which parasitize gypsy moth eggs and prevent their development the following spring. An especially imaginative cooperative program is conducted by USDA, using funds accumulated through the sales of American farm products abroad. Perhaps half of the nearly 200 major plant pests in the United States today are not native to this country. Thus USDA is financing foreign scientists to find the parasites and predators which control these pests in their native lands. Parasites for the destructive cereal leaf beetle are being sought in Poland and Yugoslavia.

• Carroll Williams' work with hormones is only a part of the research going on in an effort to alter life cycles of pestiferous plants and insects. At the University of California, scientists have used synthetic hormones to stimulate weeds to sprout in winter; tricked, the weeds respond, only to be killed off by heavy frosts. USDA scientists got similar results with hormones used on insects.

• Cultural controls, in which normal agricultural practices are modified to some extent. Crops may be planted at different times so as not to coincide with the emergence of harmful caterpillars, for instance; or they may be rotated to prevent enormous populations of a single species from building up because of its preference for a certain plant. Scientists in many lands have developed strains of plants that are resistant to insect pests and diseases; a new variety of alfalfa, for instance, exhibits some resistance to the destructive alfalfa weevil.

*

A great deal of work on these problems still faces scientists. At the moment many scientists and conservationists feel that the juvenile hormone analogues represent man's best hope to control noxious insects without blighting the rest of the environment. But effective methods of bringing the target insect into contact with the substance at the proper time must be worked out. When researchers sprayed a Colorado potato field with juvenile hormone to control the Colorado potato beetle, something went wrong with their time and dosage calculations. The result was the tardy production, at a critical stage of the season, of an army of enormous larvae that promptly ate up the potato crop.

Obviously, the utmost care must be taken when spreading chemosterilants or insect diseases in the open environment. "Natural controls" are not automatically "safe controls." Many, in fact, are far from "natural." It would be disastrous to unleash these control agents, as we did the synthetic chemical pesticides, without knowing exactly what we were doing.

Epilogue

SILENT SPRING remains a living document partly because the problems it deals with have not been solved. In light of the complexity of those problems no one (Rachel Carson least of all) expected a quick solution. What is unfortunate is that, despite the discoveries of science, the pesticide controversy goes on. Entrenched positions, dug years ago, are still hotly defended.

Yet both sides acknowledge the impact of *Silent Spring*. Picked up and amplified through government, academia, and industry, its message has brought about a number of important changes.

• Many town highway departments have reconsidered their foolish herbicidal assaults, and consequent brownouts, along country roads, restoring here and there some measure of joy to those who pass along them.

• That rain of persistent pesticides, unleashed "indiscriminately from the skies," which Rachel Carson attacked in her book, has somewhat abated; Connecticut, for example, banned the aerial spraying of DDT.

• Those massive, ill-conceived pest control programs, such as USDA's attempt to eradicate the fire ant in the Southeast, have come under closer scrutiny. Major programs in which federal agencies take part today are examined first by the Federal Committee on Pest Control.

• During the waning days of her life, Rachel Carson's concern for the nation's rivers was justified by the details made public about the Mississippi fish kill. Because of the publicity given the pesticide problem by her book, government officials and many citizens were prepared to demand effective action to prevent similar disasters.

• In the wake of the controversy stimulated by *Silent Spring*, procedures were improved by USDA for registering chemical pesticides, and by FDA for determining the maximum pesticide residues to be permitted in food and drink.

• Because of the fears she expressed about the incautious use of chemical pesticides around the home, a new attitude toward them seems to be emerging. In 1969 USDA banned the use of the cholorinated hydrocarbon, lindane, in many home vaporizors, calling them a "serious threat to human health." Soon afterward *Home Garden Magazine* took the positive step of refusing to accept advertisements for DDT and other chlorinated hydrocarbons, which "could have a harmful effect on man and his environment." The magazine urged its readers not to use these persistent pesticides. Homeowners everywhere took a new interest in pest-killers which do not contain chemical poisons; these alternate methods include black light devices which lure insects to destruction — drowning, pulverizing, or electrocuting them.

• *Silent Spring* made large areas of government and the public aware for the first time of the interrelationship of all living things and the dependence of each on a healthy environment for survival. Further, it invested the Federal Government with what Interior Secretary Udall called the "total-environment concept of conservation."

• *Silent Spring* was the beginning of that crusade which persuaded administrators and legislators alike that the chemical industry would not act in the public interest unless forced to by stricter regulations. Michigan and Arizona became the first states in the United States to ban DDT; California, beginning January 1, 1970, banned the use of DDT in home and garden. In November, 1969, the United States Government acted to phase out the use of DDT, except for "emergency uses," over a two-year period. England, as we have seen, took action against some of the most offensive chemical pesticides several years ago. Sweden announced a two-year ban on DDT, while Denmark has banned it permanently. Hungary prohibits the use of *all* chlorinated hydrocarbon pesticides. Australia, fearful that rising levels of DDT residues would cause its exported meats, produce, and dairy foods to be rejected overseas, announced late in 1968 that this chemical would be phased out as quickly as possible.*

Yet many of the problems illuminated by Rachel Carson have not been solved:

• The latest research in the United States and Great Britain gives substance to the fears she expressed for the birds of prey.

• Similar research in estuaries and in the great seas beyond them confirms her judgment that mankind, on its present course, is bringing about the "pollution of water everywhere."

• Persistent pesticides continue to appear in foods whenever scientists are curious enough to look for them. Their residues are omnipresent in fish and game. During the early months of 1969, the Food and Drug Administration seized nearly two million pounds of imported cheese in New York City: Camembert, Brie, Rouquefort, Port-Salut, Parmesan and Provolone were among the foreign cheeses returned or destroyed because of their excessive pesticide contents. Meanwhile, the Ecology Center at Berkeley, California, dramatized the buildup of DDT in mother's milk. It issued a poster, bearing the likeness of a comely, nude, and conspicuously pregnant young woman bearing the following label on her bosom: "Caution, keep out of the reach of children."

• The questions posed by Rachel Carson about the future of

* In 1969, as a part of its national drive to ban DDT, the National Audubon Society (which does not oppose the more recently developed short-lived chemical pesticides) suggested alternatives to DDT. Where some spraying is advisable, the society recommends methoxychlor, malathion, Bidrin, Diazinon, Guthion, naled, Abate, Sevin, pyrethrins, rotenone, or nicotine sulfate.

human life in a world increasingly misted with chemicals remain unanswered. As many reputable scientists caution us, we still do not know what effect pesticides (and these chemicals in combination with all the others that the body is subjected to every day) ultimately will have on human health. Final proof of safety or hazard may be some years away.

• The tide has turned dramatically against the use of DDT, yet its final disappearance must not lead our legislators and administrators to believe that the battle has been won. Dieldrin, aldrin, endrin, and the other persistent pesticides are marketed in rising quantities all over the world. Though *Silent Spring* speeded somewhat the flow of funds and the pace of research down that "other road" toward alternate controls which Rachel Carson pointed out, we have not abandoned our dependence on the most objectionable chemicals. "The problem cannot really be solved with the substances now in use," Carroll Williams wrote recently.

*

Rachel Carson was not by nature a public scold. She was happiest when preoccupied with the life of the sea. She would have preferred to be remembered for her sea books, and it is there that we will find the grandeur of style and the evocation of natural splendor that marked her work. In *Silent Spring* her reverence for life was expressed not in awe but in anger.

And so her name always will be identified with this peculiarly modern crisis. Though spokesmen for the agri-chemical industry distorted her message by implying that it led only to the surrender of the earth to noxious insects, she has been vindicated as more practical than the hard-headed businessmen. She was just as fully aware as they that insects destructive to man, his domestic animals and his crops must be controlled. She was also aware, as her critics were not, that to turn against life in all its variety a barrage of long-lasting, nonspecific poisons is as barbaric as putting a torch to the flourishing village that conceals a fugitive guerilla. Since her death, no one has stated the case more aptly than her friend, conservationist Irston R. Barnes.

"Does anyone still doubt the foresight of Rachel Carson?" he asked. "If so he need only reread *Silent Spring* and note how many voices have already been subtracted from the spring chorus, and how many more are diminished in volume. The grim irony is that a few people, in pursuit of their own narrow interests, have been able to take from all of us a part of our birthright."

References

Index

References

SINCE THIS BOOK was written for the non-specialist reader, there has been no attempt to clog the text with footnotes, or provide detailed references. The most reliable books on the subject and a list of people who contributed information to the author may be found in the Acknowledgments in the front of this book. However, for those readers interested in further background on the subject, a list of the chief sources consulted by the author in the writing of each chapter follows.

PREFACE

Page

x. C. R. Harris, Proceedings, Ontario Pollution Control Conference, Toronto, Dec. 1967.

xi. Réné Dubos, "Adapting to Pollution." *Scientist and Citizen,* Jan.–Feb. 1968.

xii. Barry Commoner, *Science and Survival.* New York, 1966.

xiv. Virginia Apgar, address to National Field Staff Conference. The National Foundation–March of Dimes, San Diego, Sept. 14, 1967.

CHAPTER 1

This and the succeeding five chapters are based in great part on the papers and correspondence (uncatalogued) of Rachel Carson, now in the possession of her literary executor, Marie Rodell of New York, N. Y. The papers eventually will be transferred to the Yale University Library. Also of great help in writing these chapters were interviews and correspondence with Shirley A. Briggs, Roland C. Clement, Clarence M. Cottam, John L. George, and Paul Knight.

3. Henry Beston, *The Freeman,* Nov. 3, 1952.

15. E. G. Hunt and A. I. Bischoff, "Inimical Effects on Wildlife of Periodic DDD Applications to Clear Lake." *California Fish and Game,* Vol. 46 (1960), No. 1.

16. Roy J. Barker, "Notes on Some Ecological Effects of DDT Sprayed on Elms." *Journal of Wildlife Management*, Vol. 22 (1958), No. 3.

CHAPTER 2

31. *Reader's Digest*, June, 1959.

CHAPTER 3

This chapter is based chiefly on conversations with, and material supplied by, Clarence M. Cottam. The reviews by Roland C. Clement appeared in *Audubon*, Nov.–Dec., 1962, and Jan.–Feb. 1964. The review by Frank E. Egler appeared in *Atlantic Naturalist*, Oct.–Dec., 1962.

CHAPTER 4

The principal reviews quoted in this chapter are by: William J. Darby (*Chemical and Engineering News*, Oct. 1, 1962); Clarence M. Cottam (*Sierra Club Bulletin*, Jan., 1963); LaMont C. Cole (*Scientific American*, Dec., 1962); Loren C. Eiseley (*Saturday Review*, Sept. 29, 1962); and Robert L. Rudd (*Pacific Discovery*, Nov.–Dec. 1962).

CHAPTER 5

Much valuable information for this chapter was supplied to the author by Dr. William H. Drury, Jr., a member of the Panel on the Use of Pesticides of the President's Science Advisory Committee, and Paul Knight, a member of the Secretary of the Interior's Program Support Staff.

69. *Time*, Sept. 28, 1962.
70. Roland C. Clement, "Education in the Use of Pesticides." An address to the Natural Resources Conference (sponsored by the Garden Club Federation of Pennsylvania), March 10, 1964.
73. Sylvia Porter's syndicated newspaper column (in *New York Post*, etc.), April 20, 1964.
79. *Chemical and Engineering News*, May 20, 1963.
79. *Science*, May 24, 1963.

CHAPTER 6

Much of the information on events in Great Britain were supplied to the author by Lord Shackleton, Stanley Cramp, N. W. Moore, and Peter Scott. C. J. Briejèr's letters to Rachel Carson and the author were also invaluable.

81. Julian Huxley, letter to *New York Times*, April 19, 1964.
86. Kenneth Mellanby, *Pesticides and Pollution.* London, 1967.

CHAPTER 7

This chapter is based largely on the *Transcript of the Conference on the Pollution of Interstate Waters* (*Lower Mississippi River*), conducted by the United States Public Health Service at New Orleans, May 5–6, 1964. Donald I. Mount and Murray Stein, formerly of the Public Health Service and now with the Federal Water Pollution Control Administration, both supplied helpful additional material. The author also treated this event in a slightly different context in his book, *Disaster By Default*, New York, 1966.

94. Barry Commoner, *Science and Survival.* New York, 1966.
100. Lucille F. Stickel, "Organochlorine Pesticides in the Environment." Mimeographed. Patuxent Wildlife Research Center, Laurel, Md., 1968.
104. Secretary of the Interior, Memorandum to Heads of Bureaus and Offices, May 7, 1964.
106. *Pesticides and Public Policy:* Report of the Committee on Government Operations. Subcommittee on Reorganization and Internal Organization, U.S. Senate, July 21, 1966 (the so-called Ribicoff Report).

CHAPTERS 8 AND 9

Because these chapters deal with the latest evidence against the chlorinated hydrocarbon pesticides which has vindicated Rachel Carson's *Silent Spring*, a more detailed list of references is provided here. This bibliography was compiled by, and is used with the permission of, Charles F. Wurster, Jr.

ABBOTT, D. C., HARRISON, R. B., TATTON, J. O'G., & THOMSON, J. (1965). Organochlorine pesticides in the atmospheric environment. Nature, *208*, 1317–8.

AMES, P. L. (1966). DDT residues in the eggs of the Osprey in the northeastern United States and their relation to nesting success. J. Appl. Ecol. *3* (Suppl.), 87–97.

BOWMAN, M. C., ACREE, F., & CORBETT, M. K. (1960). Solubility of carbon-14 DDT in water. J. Agr. Food Chem. *8*, 406–8.

BOWMAN, M. C., ACREE, F., LOFGREN, C. S., & BEROZA, M. (1964). Chlorinated insecticides: fate in aqueous suspensions containing mosquito larvae. Science *146*, 1480–1.

BURDICK, G. E., et al. (1964). The accumulation of DDT in lake trout and the effect on reproduction. Trans. Am. Fish. Soc. *93*, 127–36.

BUTLER, P. A., & SPRINGER, P. F. (1963). Pesticides — a new factor in coastal environments. Trans. 28th North Amer. Wildl. Nat. Res. Conf., Mar. 4–6, 378–90.

BUTLER, P. A. (1966). Pesticides in the marine environment. J. Appl. Ecol. *3* (Suppl.), 253–9.

CAIRNS, J., & SCHEIER, A. (1964). The effect upon the pumpkinseed sunfish *Lepomis gibbosus* (Linn.) of chronic exposure to lethal and sublethal concentrations of dieldrin. Notulae Naturae, No. 370, 1–10.

CHANT, D. A. (1966). Integrated control systems. Nat. Acad. Sci. Nat. Res. Council Publ. 1402, 193–218.

COLE, H., BARRY, D., FREAR, D. E. H., & BRADFORD, A. (1967). DDT levels in fish, streams, stream sediments, and soil before and after aerial spray application for fall cankerworm in northern Pennsylvania. Bull. Environ. Contam. Toxicol. *2*, 127–46.

CONNEY, A. H. (1967). Pharmacological implications of microsomal enzyme induction. Pharmacol. Rev. *19*, 317–66.

EDWARDS, C. A. (1966). Insecticide residues in soils. Residue Rev. *13*, 83–132.

ELSON, P. F. (1967). Effects on wild young salmon of spraying DDT over New Brunswick forests. J. Fish. Res. Bd. Canada *24*, 731–67.

GROSCH, D. S. (1967). Poisoning with DDT: effect on reproductive performance of Artemia. Science *155*, 592–3.

GUENZI, W. D., & BEARD, W. E. (1967). Anaerobic biodegradation of DDT to DDD in soil. Science *156*, 1116–7.

HERBERT, R. A., & HERBERT, K. G. S. (1965). Behavior of Peregrine Falcons in the New York City region. Auk *82*, 62–94.

HERMAN, S. G., GARRETT, R. L., & RUDD, R. L. (1968). Pesticides and the Western Grebe. First Rochester Conf. Toxicity, Univ. of Rochester, 4–6 June 1968.

HICKEY, J. J., KEITH, J. A., & COON, F. B. (1966). An exploration of pesticides in a Lake Michigan ecosystem. J. Appl. Ecol. *3* (Suppl.), 141–54.

HICKEY, J. J., Ed. (1968). Peregrine Falcon populations, their biology and decline. Univ. of Wisc. Press, Madison.

HUNT, L. B. (1960). Songbird breeding populations in DDT-sprayed Dutch elm disease communities. J. Wildl. Mgmt. *24*, 139–46.

IDE, F. P. (1967). Effects of forest spraying with DDT on aquatic insects of salmon streams in New Brunswick. J. Fish. Res. Bd. Canada *24*, 769–805.

JOHNSON, H. (1968). Press release, Mich. State Univ., Mar. 7.

KEITH, J. A. (1966). Reproductive success in a DDT-contaminated population of Herring Gulls. J. Appl. Ecol. *3* (Suppl.), 57–70.

KEITH, J. O., & HUNT, E. G. (1966). Levels of insecticide residues in fish and wildlife in California. Trans. 31st North Amer. Wildl. Nat. Res. Conf., Mar. 14–16, 150–77.

KUPFER, D. (1967). Effects of some pesticides and related compounds on steroid function and metabolism. Residue Rev. *19*, 11–30.

MACEK, K. J. (1968). Reproduction in brook trout (*Salvelinus frontinalis*) fed sublethal concentrations of DDT. J. Fish. Res. Bd. Canada *25*, 1787–96. (Also see 2443–51.)

MATTHYSSE, J. G. (1959). An evaluation of mist blowing and sanitation in Dutch elm disease control programs. Cornell Misc. Bull. 30, N. Y. State Coll. Agr., Ithaca, N. Y., 16 pages.

NASH, R. G., & WOOLSON, E. A. (1967). Persistence of chlorinated hydrocarbon insecticides in soils. Science *157*, 924–7.

O'BRIEN, R. D. (1967). Insecticides, action and metabolism. Academic Press, N. Y., 332 pages.

OGILVIE, D. M., & ANDERSON, J. M. (1965). Effect of DDT on temperature selection by young Atlantic salmon, *Salmo salar*. J. Fish. Res. Bd. Canada *22*, 503–12.

PEAKALL, D. B. (1967). Pesticide-induced enzyme breakdown of steroids in birds. Nature *216*, 505–6.

PORTER, R. D., & WIEMEYER, S. N. (1969). Dieldrin and DDT: effects on sparrow hawk eggshells and reproduction. Science *165*, 199–200.

QUINBY, G. E., HAYES, W. J., ARMSTRONG, J. F., & DURHAM, W. F. (1965). DDT storage in the U. S. population. J. Am. Med. Assoc. *191*, 175–9.

RATCLIFFE, D. A. (1967). The peregrine situation in Great Britain 1965–66. Bird Study *14*, 238–46.

RATCLIFFE, D. A. (1967). Decrease in eggshell weight in certain birds of prey. Nature *215*, 208–10.

RISEBROUGH, R. W. (1968). Chlorinated hydrocarbons in marine ecosystems. First Rochester Conf. Toxicity, Univ. of Rochester, 4–6 June.

RISEBROUGH, R. W., HUGGETT, R. J., GRIFFIN, J. J., & GOLDBERG, E. D. (1968). Pesticides: transatlantic movements in the northeast trades. Science *159*, 1233–6.

SIMKISS, K. (1961). Calcium metabolism and avian reproduction. Biol. Rev. *36*, 321–67.

SLADEN, W. J. L., MENZIE, C. M., & REICHEL, W. L. (1966). DDT residues in Adelie penguins and a crabeater seal from Antarctica: ecological implications. Nature *210*, 670–3.

WARNER, R. E., PETERSON, K. K., & BORGMAN, L. (1966). Behavioral pathology in fish: a quantitative study of sublethal pesticide toxication. J. Appl. Ecol. *3* (Suppl.), 223–47.

WOODWELL, G. M. (1961). The persistence of DDT in a forest soil. Forest Sci. 7, 194–6.

WOODWELL, G. M., WURSTER, C. F., & ISAACSON, P. A. (1967). DDT residues in an East Coast estuary: a case of biological concentration of a persistent insecticide. Science *156*, 821–4.

WRIGHT, B. S. (1965). Some effects of heptachlor and DDT on New Brunswick woodcocks. J. Wildl. Mgmt. *29*, 172–85.

WURSTER, D. H., WURSTER, C. F., & STRICKLAND, W. N. (1965). Bird mortality following DDT spray for Dutch elm disease. Ecology *46*, 488–99.

WURSTER, C. F., & WINGATE, D. B. (1968). DDT residues and declining reproduction in the Bermuda Petrel. Science *159*, 979–81.

WURSTER, C. F. (1968). DDT reduces photosynthesis by marine phytoplankton. Science *159*, 1474–5.

WURSTER, C. F. (1968). Chlorinated hydrocarbon insecticides and avian reproduction: how are they related? First Rochester Conf. Toxicity, Univ. of Rochester, June 4–6.

CHAPTER 10

Among the chief sources for this chapter: D. E. Weeks, "Endrin Food Poisoning." *Bulletin of the World Health Organization*, Vol. 37 (1967), No. 4; G. A. Reich, "Pesticides and Public Health." *Proceedings of the Seminars on Pesticides in Our Environment*, Univ. of Maine, Oct. 17–18, 1967; Clyde M. Berry, "Industrialization of Agriculture," *Atlantic Naturalist*, Jan.–March, 1962; *Toxic Hazards in Aerial Application*, Federal Aviation Agency, April, 1962; *Central Nervous System Effects of Chronic Exposure to Organophosphate Insecticides*. Federal Aviation Agency, Oct., 1963; *Briefs of Accidents Involving Aerial Application Operations*, Civil Aeronautics Board, 1964; G. M. Woodwell, "Toxic Substances and Ecological Cycles." *Scientific American*, March, 1967; Irma West, "Detection and Control of Pesticide Poisoning." *First Rochester Conference on Toxicity*, Univ. of Rochester, June 4–6, 1968; Lamont C. Cole, *Scientific American*, Dec., 1962; William H. Drury, Jr., interviews with author.

CHAPTER 11

143. Réné Dubos, "Adapting to Pollution." *Scientist and Citizen*, Jan.–Feb., 1968.

144. Morton Mintz, *Washington Post*, April 24, 1968.

144. Clair C. Patterson, *Hearings before a Subcommittee on Air and Water Pollution of the Committee on Public Works*. U.S. Senate, June 7, 8, 9, 14 and 15, 1966.

145. Louis Lasagna, "The Diseases Drugs Cause." *Perspectives in Biology and Medicine*, Summer, 1964.

145. *Audubon Leader's Conservation Guide*, Dec. 15, 1967.

146. Bill Kovach, *New York Times*, July 10, 1969.

147. *Medical World News*, March 14, 1969.

148. Victor Cohn, *Washington Post*, April 28, 1968.

149. *Medical World News*, April 26, 1968.

149. Virginia Apgar, address to National Field Staff Conference. The National Foundation–March of Dimes, San Diego, Sept. 14, 1967.

150. Göran Löfroth with Margaret E. Duffy, "Birds Give Warning." *Environment*, May, 1969.

CHAPTER 12

152. J. Headly and J. Lewis, *The Pesticide Problem: An Economic Approach to Public Policy*. Resources for the Future, Washington, 1967.

152. Roland C. Clement, "The Pesticide Problem." *National Resources Journal*, January, 1968.

155. Kevin P. Shea, "Cotton and Chemicals." *Scientist and Citizen*, Nov., 1968.

156. "Safe Use of Pesticides in Public Health." World Health Organization Technical Report Series, No. 356, Geneva, 1967.

157–59. Based on material supplied to the author by Charles F. Wurster, Jr., and N. W. Moore.

159. G. M. Woodwell, *Radioactivity and Fallout: The Model Pollution*. Mimeographed. Brookhaven National Laboratory, 1968.

CHAPTER 13

166. Barry Commoner, in *Report on Session III, Symposium on Secrecy, Privacy and Public Information*. AAAS, Annual Meeting, Dec., 1967.

166. G. G. Simpson, "Biology and the Public Good." *American Scientist*, June, 1967.

167. Philip Abelson, "Are the Tame Cats in Charge?" *Saturday Review*, Jan. 1, 1966.

167–68. Based on material supplied to the author by Robert L. Rudd.

171. Frank E. Egler, "Pesticides in Our Ecosystem." *BioScience*, Nov. 1964.

171–72. L. W. Boulanger, "Pest Control, Past, Present and Future." *Proceedings of the Seminars on Pesticides in Our Environment*, Univ. of Maine, Oct. 17–18, 1967.

173–74. Kevin P. Shea, "Cotton and Chemicals," *Scientist and Citizen*, Nov., 1968.

CHAPTER 14

This chapter is based chiefly on the following publications:

Effects, Uses, Control, and Research of Agricultural Pesticides, A Report by the Surveys and Investigations Staff, Subcommittee on Department of Agriculture of the Committee on Appropriations, House of Representatives, Washington, 1966.

Jamie L. Whitten, *That We May Live*. Princeton, N. J., 1966.

CHAPTER 15

This chapter is based chiefly on interviews with a number of government officials in Washington.

CHAPTER 16

193–94. G. G. Simpson, "Biology and the Public Good." *American Scientist*, June, 1967.
197. Richard M. Leonard, telegram to the Secretary of the Interior, June 10, 1963.
200–01. Advisory Board on Wildlife Management, *Report on Predator and Rodent Control.* Report to the Secretary of the Interior, March, 1964.
202–03. John Madson, "Dark Days in Dogtown." *Audubon*, Jan.–Feb., 1968.

CHAPTER 17

207. George R. Harvey and Jay D. Mann, "Picloram in Vietnam." *Scientist and Citizen*, Sept., 1968.
207. Walter Sullivan, *New York Times*, Feb. 13, 1968.

CHAPTER 18

This chapter is based chiefly on the following article:

Shirley A. Briggs, "The *Aedes aegypti* Eradication Program." *Atlantic Naturalist*, Oct.–Dec., 1965.

CHAPTER 19

This chapter is based chiefly on the following sources: Material supplied to the author by S. C. Billings of USDA, Hon. Catherine May, and Robert L. Rudd; and various publications of the U.S. Department of Agriculture, including its *Annual Review*.

225. Frank E. Egler, "Pesticides in our Ecosystem." *American Scientist*, March, 1964.
227. LaMont C. Cole, "The Complexity of Pest Control in the Environment." *Scientific Aspects of Pest Control*, NAS–NCR, Washington, 1966.
233–34. *New York Times*, Sept. 17, 1968.

CHAPTER 20

This chapter is based chiefly on material supplied to the author by Dale F. Bray, Shirley A. Briggs, Roland C. Clement, Stephen Collins, Wayne Hanley, Alfred L. Hawkes, Allen Morgan, Robert L. Rudd, F. J. Trembley, and G. M. Woodwell.

240. Louis J. Halle, "Men and Birds: The Changing Relationship." *Atlantic Naturalist*, Summer, 1969.

242. Philip M. Boffey, "Du Pont and Delaware: Academic Life Behind the Nylon Curtain." *Science*, May 10, 1968.

CHAPTER 21

This chapter is based chiefly on material supplied to the author by Roland C. Clement, Charles F. Wurster, Jr., and Victor J. Yannacone, Jr.

258. *Agrichemical West*, April, 1968.
258–59. *Farm Chemicals*, Jan., 1968.

CHAPTER 22

This chapter is based chiefly on the following articles:

Carroll M. Williams, "Third-Generation Pesticides." *Scientific American*, July, 1967.

Lawrence Lessing, "A Molecular Bomb for the War Against Insects." *Fortune*, July, 1968.

EPILOGUE

271. Irston R. Barnes, "Perspectives." *Atlantic Naturalist*, Summer, 1969.

Index